Science in the Looking Glass

Science in the Looking Glass

What Do Scientists Really Know?

E. Brian Davies

Department of Mathematics
King's College London

OXFORD
UNIVERSITY PRESS

OXFORD

UNIVERSITY PRESS

Great Clarendon Street, Oxford OX2 6DP

Oxford University Press is a department of the University of Oxford.
It furthers the University's objective of excellence in research, scholarship,
and education by publishing worldwide in

Oxford New York

Auckland Bangkok Buenos Aires Cape Town Chennai
Dar es Salaam Delhi Hong Kong Istanbul Karachi Kolkata
Kuala Lumpur Madrid Melbourne Mexico City Mumbai Nairobi
São Paulo Shanghai Taipei Tokyo Toronto

Oxford is a registered trade mark of Oxford University Press
in the UK and in certain other countries

Published in the United States
by Oxford University Press Inc., New York

A catalogue record for this title is available from the British Library

Library of Congress Cataloging in Publication Data
(Data available)
ISBN 0 19 852543 5

10 9 8 7 6 5 4 3 2 1

Typeset by Newgen Imaging Systems (P) Ltd., Chennai, India
Printed in Great Britain
on acid-free paper by
Biddles Ltd, www.biddles.co.uk

Preface

Almost every month some book or television programme describes exciting developments in cosmology or fundamental physics. Many tell us that we are on the verge of finding the explanation for the Big Bang or the ultimate Theory of Everything. These will explain all physics in one fundamental set of mathematical equations. It is easy to be swept along by the obvious enthusiasm of the participants, particularly when they are making real progress in pushing back the boundaries of knowledge. Unfortunately, most of their brilliant new ideas are doomed to be forgotten, if only because they cannot all be right.

Consider the currently fashionable idea that our universe is just one of many unobservable, parallel universes, all equally real. How can one hope to describe the inner structures of such universes, each with its own values of the 'fundamental' constants? Many may be dull and featureless, but others are presumably as fascinating and complex as our own. However much some physicists declare the reality of these other universes, in practice their main function is to support the mathematical models of the day, or to 'explain' certain properties of our own universe.

My goal in this book is not to adjudicate on the correctness of such new and speculative theories. We will instead consider the development of science in a historical context, in order to find out how such questions have been resolved in the past, and to explain why many long established 'facts' have turned out not to be so certain. My conclusion is surprising, particularly coming from a mathematician. In spite of the fact that highly mathematical theories often provide very accurate predictions, we should not, on that account, think that such theories are true or that Nature is governed by mathematics. In fact the scientific theories most likely to be around in a thousand years' time are those which are the *least* mathematical—for example evolution, plate tectonics, and the existence of atoms.

The entire book is effectively an extended defence of the above statements. In the course of the discussion I risk the displeasure of many of my colleagues by explaining the feebleness of mathematical Platonism as a philosophy. I also provide psychological and historical support for the claim that mathematics is a human creation. Its success in explaining nature is a result of the fact that we developed much of it for precisely that purpose. Even the numbers which we use in counting become no more than formal symbols, invented by us, as soon as they are as big as 10^{1000} (1 followed by a thousand zeros). Pretending that we can count from 1 up to such a number 'in principle' is a fantasy, and will always remain so. Moreover, it is not necessary to believe this in order to be interested in pure mathematics.

Whatever some over-enthusiastic physicists might claim, there is much which is beyond our grasp, and which will probably remain so. Subjective (first person) consciousness is one such issue. Understanding the true nature of quantum particles is another, in spite of the proven success of the mathematical aspects of quantum theory. Contingency, or historical accident, has obviously had a major influence on geology and biology, but some physicists think that it is even involved in the form of the laws of physics. Whether or not this is true, scientists are right to believe that, with enough effort, they can push the boundaries of their subjects far beyond their present limits.

An unusual feature of the book is that I try to explain why philosophical issues are important in science by means of simple examples. This is not the style followed by academic philosophers, but it makes the issues easier to understand, particularly in a popular context. In addition, discussions about the status of money, zombies, or rainbows are more fun than dry logical arguments about ontology.

I am painfully aware that the scope of the book is far wider than anybody's expertise could span in this age of specialists. The attempt is worth making, because arguments informed by only one branch of science are inevitably distorted by that fact. I do not claim to have found the final answer to all of the deep questions in the philosophy of science, but hope that readers who have not previously thought much about these will see why they are important.

People vary enormously in their liking of mathematics. Many switch off as soon as they see it, and editors of popular books advise their authors to reduce it to the absolute minimum. I have gone as far as I can in this direction, and reassure the allergic reader that any difficult passages can be skimmed over. They are present to ensure that interested readers do not feel cheated by being told conclusions without any evidence in their support.

I wish to acknowledge invaluable advice, or sometimes just stimulation, which I have received from many friends and colleagues, in particular Martin Berry, Alan Cook, Richard Davies, Donald Gillies, Nicholas Green, Andreas Hinz, Hubert Kalf, Mike Lambrou, Peter Palmer, Roger Penrose, David Robinson, Peter Saunders, Ray Streater, John Taylor and Phil Whitfield. I do not, however, burden them with the responsibility of agreeing with anything I say here. I also thank my family for providing an atmosphere in which a task such as this could be contemplated; I know that the time which I have devoted to it has put me in great debt to them.

Contents

—————————

1
Perception and Language

1.1 Preamble

Most of the time most people relate to the world in a pretty straightforward way. We assume that entities which appear to exist actually do so, and expect scientists to provide us with steadily more detailed descriptions of their underlying structures. We try not to worry about the fact that fundamental theories are highly mathematical, and hence incomprehensible to almost everyone. Some, such as the Oxford chemist Peter Atkins, find the prospect of ultimately explaining the whole of reality in mathematical terms exhilarating, while others fear or reject it because of its impersonal character.

There are a few puzzles associated with this scientific picture of reality. One is the nature of subjective consciousness, which used to be called the human soul, and which some philosophers now regard as an illusion. Another is the status of mathematics: why should the ultimate explanation of reality be in terms of equations?

Roger Penrose has addressed these fundamental questions in his books *The Emperor's New Mind* and *Shadows of the Mind*, published in 1989 and 1994 respectively. Roger is an outstanding mathematical physicist, but I think that his approach to these issues is quite wrong, and in this book I propose an entirely different way of looking at them. Readers will probably be relieved to hear that they are not going to be asked to wade through page after page of detailed mathematics or logic. Although it contains some mathematical results as illustrations, this book does not involve any deep technical arguments.

One of Penrose's principal ideas is that Gödel's theorems, discussed on page 111, prove that human beings can understand results which are beyond the capacity of any computer. He believes that they also provide a route by means of which one can understand the mathematical mind, and by extension the nature of consciousness. This is pretty optimistic, to say the least. Penrose makes strong statements about the limitations of computers, but ignores the obvious fact that the human mind also has limits.

Mathematics provides one of the last refuges of Platonism, discussed in some detail on pages 27 and 37. I will argue that this philosophy is entirely unhelpful in understanding either mathematics or its relationship with the outside world. The high degree of abstractness of the subject is shared by chess, philosophy, and music, and does not require any special explanation. I thus reject the Platonistic position of a sizeable fraction of my colleagues, including some of the most eminent. On the other hand, the ideas presented here are entirely in line with modern experimental psychology and the history of mathematics itself.

Re-establishing the links between mathematics, science, and other human concerns involves a rejection of the 'easy' reductionist option, which leaves subjective consciousness out in the cold. This book does not provide the solution to every problem about the nature of reality, but presents a series of arguments suggesting that we must stop looking in directions which leave us out of the picture. Platonism, in which mathematics exists in some ideal world unrelated to human society, is a typical example of this. Since the time of Descartes, Western science has developed along a route which has been immensely successful for those aspects of reality in which human issues are of little relevance. Its very success has encouraged scientists to avert their gaze from those aspects of reality which their methods say little about. Some have even convinced themselves that there are no such aspects.

In this chapter we consider the evidence that *almost everything* relating to human knowledge is more problematical that we normally admit. We start with a review of recent work in experimental psychology, because it is surely necessary to understand our physical nature if we are to understand the nature of our thoughts. This chapter is absolutely mainstream psychology. I cannot make quite the same claim about Chapter 2, because most deep questions in philosophy remain controversial. From Chapter 3 onwards we will cover a wide range of sciences, indeed any area in which there is controversy about the bases for claims of objective knowledge.

The first half of this chapter describes the wide variety of methods which have been used to investigate the differences between what we think we see and reality itself. Not only have these investigations provided a consistent description of the world, but they even explain why our unaided senses paint a distorted, indeed different, picture. Particularly important in this respect has been the development of brain scanning machines, which are beginning to give detailed information about what is happening as our brains struggle to interpret sensory data. This is one of the most exciting current fields of scientific research.

As a society we are progressively re-adjusting our world-view in the direction indicated by our instruments and intellects. To give just one example: we commonly talk about a 'fluid' called electricity which can flow through solid copper wires but not through the open air; this fluid can be stored in batteries, even though a full battery looks the same and is no heavier than an empty one.[1] We accept such bizarre propositions in spite of a complete lack of direct sensory evidence because they provide consistent explanations of observed phenomena, such as the fact that a light bulb becomes bright when we turn a switch. For the first time in history large parts of our lives depend upon machines and ideas which would appear magical or incomprehensible to our ancestors.

In the second half of the chapter we discuss the relationship between language and reality, which turns out to be a much harder task.

1.2   Light and Vision

Introduction

The view that our senses provide us with direct and straightforward information about the outside world was promulgated by Aristotle, St. Thomas Aquinas, and then by the sixteenth century scholastic philosophers. The first person to criticize it systematically was Descartes, whose philosophical and scientific ideas will be discussed in more detail in Chapter 2. In *Le Monde*, 1632 he wrote:

> *In proposing to treat here of light, the first thing I want to make clear to you is that there can be a difference between our sensation of light . . . and what is in the objects that produces that sensation in us . . . For, even though everyone is commonly persuaded that the ideas that are the objects of our thought are wholly like the objects from which they proceed, I see no reasoning that assures us that this is the case.*

Newton later provided positive reasons, described below, for distinguishing between colours and our sensations of them, and these have been reinforced by all recent psychological research. Our present understanding of brain function has involved many different lines of investigation. One is the study of optical illusions, which provide hints about the brain mechanisms involved in 'normal' vision. Secondly, psychologists study the abnormal thought processes of people who have suffered specific brain damage; this helps them to discover which regions of the brain are involved in different types of processing. There has been extensive analysis of the biochemistry and structure of individual nerve cells, and of the anatomy of the retina and the rest of the brain. Another rapidly developing field of psychological research depends upon the use of brain scanning machines: these can identify which parts of the brain are most active when people are asked to carry out various mental tasks. Research in each of these fields forces us to the conclusion that the unconscious part of our brain constructs the reality in which we live; evolution has seen to it that these mental constructions lead to appropriate behaviour in most normal circumstances. Donald Hoffman gives a clear statement of this conclusion from the point of view of an experimental psychologist in *Visual Intelligence: How We Create What We See*. He explains why it is possible for us all to agree about the nature of the world and nevertheless for us to be fundamentally wrong in the way we see it.

> *Subjective pictures are not just part of picture perception. They are part of ordinary everyday seeing. And that should come as no surprise. You construct*

every figure you see. So, in this sense, every figure you see is subjective....
But then why do we all see the same thing? Is the consensus magic? No. We
have consensus because we all have the same rules of construction.

According to Hoffman the rules of construction are built into the anatomy of
our brains, and cannot be modified by the exercise of rational thought. Lest you
think that this is just Hoffman's personal view, let me quote a corresponding
passage from Francis Crick's *The Astonishing Hypothesis*.

What you see is not what is really there; it is what your brain believes is there. In
many cases this will indeed correspond well with characteristics of the visual
world before you, but in some cases your 'beliefs' may be wrong. Seeing
is an active constructive process. You brain makes the best interpretation
it can according to its previous experience and the limited and ambiguous
information provided by your eyes.

These ideas seem rather disturbing, but would have been regarded as absolutely
orthodox Taoist philosophy in tenth century China. The book *Hua Shu* of this
period describes a kind of subjective realism, in which the external world is real,
but our knowledge of it is deeply affected by the way in which it is perceived,
so that we cannot seize its full reality. Like Hoffman and Crick, the (supposed)
author T'an Ch'iao even refers to optical illusions and human inattention to press
the view that we pick out certain elements of reality to form our world-picture.[2]

The ideas above provide strong warnings against believing that something
is true simply because it matches our intuition well. We *can* gain objective
knowledge about the underlying reality, but this depends upon learning to accept
the verdict of our instruments rather than of our unaided senses. We have chosen
this path because such a wide variety of different methods of scientific investigation have led to a consistent picture. Indeed they even explain *why* the evidence
of our own senses is not a reliable guide to the nature of reality.

The Perception of Colour

The study of optical phenomena was slow to develop historically because of the
great difficulty of disentangling the physical, physiological, and psychological
aspects of the subject. It provides a very clear example of the immense gap
between our perceptions and the physical reality which lies behind them.

Although the Pythagoreans maintained that light travelled from the eye to
the object, Lucretius got much closer to the truth in *The Nature of the Universe*:

No matter how suddenly or at what time you set any object in front of a mirror,
an image appears. From this you may infer that the surfaces of objects emit
a ceaseless stream of flimsy tissues and filmy shapes. Therefore a great many
films are generated in a brief space of time, so that their origins may rightly
be described as instantaneous. Just as a great many particles of light must
be emitted in a brief space of time by the sun to keep the world continually
filled with it, so objects in general must correspondingly send off a great
many images in a great many ways from every surface and in all directions
simultaneously.

Lucretius also argued that the colours of objects were not intrinsic to them, since the sea could have a variety of appearances according to the way that its component atoms were churning around inside it. In spite of the startling accuracy of these ideas for the time, they cannot be classified as science: Lucretius could propose no quantitative tests of his ideas, which were, eventually, just speculation.

The scientific investigation of light and colour started in the seventeenth century, with major investigations by Robert Hooke, Christiaan Huygens, and Isaac Newton. Newton used prisms to split white light into its component colours, and then recombined the components back into white light; this led him to understand that white light was not pure, as seems naively obvious, but a mixture. He understood the cause of chromatic aberration in the lenses of refracting telescopes, and designed and built the first reflecting telescope in 1669. His paper *Theory of Light and Colours*, published in 1672, attracted great attention and also started a feud between him and Hooke. They differed sharply about whether light should be regarded as corpuscular or wave-like, a debate which was to continue until the twentieth century, when quantum mechanics allowed it to be both.

We now know that light comes in a continuous range of wavelengths, and that our eyes are only sensitive to a very narrow band of these. Our colour discrimination depends upon our having three kinds of receptor, called R, G, and B cones, in our retinas, each of which is most sensitive to a particular range of wavelengths. These receptors cannot possibly distinguish between all the wavelengths in visible light, so what we see is a great simplification of what is in the light itself. Objects are *not* red, green, or blue in themselves: our impressions are created by neural processing of the very limited information provided by our retinas.

People actually have one of two types of R cones, which are genetically inherited. These produce slight differences in perception between individuals, which may be important when matching colours. The variation is caused by a single amino acid change on the relevant gene, and provides a rare instance in which we know the precise causal chain from a change at the molecular level to a difference between the subjective worlds two people may inhabit.[3] This provides a partial answer to the philosophical question of how we can know that two normal people have the same subjective colour experiences: they need not.

This is not merely an abstract problem. I myself have had regular disagreements over many years with my wife about the nature of colours on the borderline between green and blue. We cannot even agree whether this is a difference of naming or of perception. Maybe our colour receptors are indeed slightly different, and we have been taught to name colours by parents who had the same types of receptor as ourselves.

These small differences pale into insignificance once one compares our visual experiences with those of other species. It is known that many birds and insects are sensitive to ultraviolet light. Ultraviolet photographs of some flowers reveal patterns, invisible to us, which are important to insects seeking nectar.

On the other hand most mammals have only two kinds of colour receptor. Our very similar R and G cones appear to have evolved from an earlier single type recently, possibly to improve our ability to discriminate between fruits.[4] In the most common form of colour blindness, affecting very roughly 5% of males, either the R or the G cones are missing, so the person cannot distinguish red from green. We regard such people as having a disability, but by the standards of most mammals they are normal. On the other hand pigeons have six or more different types of colour receptor, and might regard all humans as having only partial colour vision! We conclude that the same light falling on the eyes of different species must produce very different subjective colour impressions in their brains.

Returning to human beings, it is known that quite different combinations of wavelengths may produce the same subjective impression. Whether the *names* of colours are simply cultural constructs or have a physiological basis is again a matter of active debate. The comparative study of a large number of languages shows that although they may have different numbers of named colours these are classified into a coherent hierarchy. Namely if any colour in the box below appears in a language then all of the colours on previous lines also appear.

This suggests that there is a physiological basis for the existence of colour names, even if there is no external physical basis. Unfortunately, even this conclusion has recently been thrown into doubt by a study of the Burinmo tribe in Papua New Guinea. The colour names of this tribe are radically different from the above list and their ability to distinguish colours is positively correlated to their colour language. These observations do not support the idea that colour categories could be universal.[5] In biology almost everything is more complicated than initial analyses suggest!

> white, black
> red
> green, yellow
> blue
> brown
> purple, pink, etc.

Interpretation and Illusion

There are many other differences between our perceptions and the reality behind them. When light falls on a retinal receptor, it emits pulses which are then processed in stages, first in the retina and then in the brain. Each level of processing involves further interpretation and selection, all of which happens before we become conscious of the scene before us. In most cases we are unaware that these processes are going on, but it is possible to set up situations in which we can see that our lower level interpretations are quite incorrect. Understanding the way in which images are processed is a research field of great complexity, and my goal here is simply to draw attention to the variety of mechanisms involved, and a few of the ways in which they can fail. When this happens we experience an optical illusion.

A simple example, much exploited by Bridget Riley and other artists in the Op Art movement of the 1960s, involves a property of our peripheral vision. The phenomenon can be seen by moving your head towards and away from figure 1.1, while concentrating on the spot in the centre. The strong sense that the rings are rotating, in opposite directions, depends upon the fact that the peripheral part of our retina is primarily concerned with the detection of movement. The neural circuits involved are designed, for obvious evolutionary reasons, to 'fail-safe': it does not matter too much if a non-existent movement is reported, but might be fatal if an actual movement is missed, even once. Even when we recognize that the effect is illusory, we cannot prevent it happening, because the neural processing happens below the level at which we have conscious control.

Judging the brightness of a part of a picture is not a simple matter. In figure 1.2, drawn by Ted Adelson, the two squares labelled A and B are exactly the same shade of grey. This may be checked by covering up everything in the picture except these two squares. The reason for the illusion is that your visual system is not interested in the true luminosity of the squares. One part interprets the picture as being of a three-dimensional object, and passes this conclusion to

Fig. 1.1 Rotating Rings

Fig. 1.2 Checker-shadow Illusion
Reproduced by permission of Edward H. Adelson, Department of Brain and Cognitive
Science, MIT

another part, which compensates for what it considers to be the likely variations
of lighting. By the time you become conscious of the picture, these adjustments
are simply a part of what you see.

The extent to which 'seeing' depends upon active brain processes became
very clear to me on a recent holiday in Madeira. Standing on the edge of a
shallow pool one sunny day, a companion remarked on the number of small
fish in it. Although I looked hard through the constantly varying pattern of
surface ripples I could not see any. My companion explained carefully what
I should look for and within a minute or so my brain reprogrammed itself, and
hundreds of the fish became clearly visible. Indeed I could hardly understand
how I had not been able to see them before.

This is not an isolated example. So-called 'primitive' peoples learn to recognize myriads of subtle features of their environments which urban travellers are
completely unaware of. These may be vital for avoiding dangers as well as
for finding sources of food. Figure 1.3 'Random Points' shows how powerful
the mechanisms involved are. As soon as you are told that there is one 'extra'
point in the figure, you can identify it as one out of two possibilities without
consciously looking at most of them. This feat can only be achieved so quickly
because your visual system processes the whole picture simultaneously. In
computer terms it is a massively parallel system. Fortunately such tasks do not

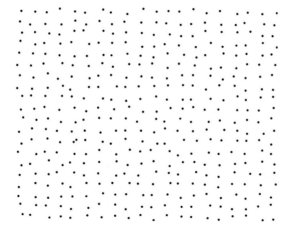

Fig. 1.3 Random Points

need to be carried out using our rational faculties, which would be much less competent at such tasks!

Let us turn to the way in which we construct three-dimensionality from what we see. Following the re-discovery and elucidation of the laws of perspective by Brunelleschi and Alberti in the first half of the fifteenth century, Hogarth was one of the first painters to produce pictures with deliberately impossible perspectives. In fact it is embarrassingly easy to follow the laws of perspective rigorously, while producing impossible objects, as figure 1.4 shows.

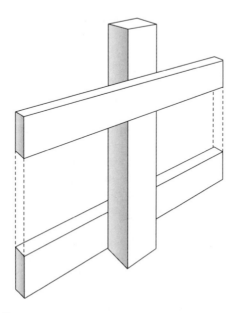

Fig. 1.4 Part of a Fence

Similar ideas underlie several of the paintings of M. C. Escher, such as *Ascending and Descending*, 1960, in which a chain of monks climb a staircase which apparently returns to its starting point, even though every step is upwards. Escher cleverly incorporated enough distractions into the picture that it does not appear particularly strange to the eye. These illusions are possible because our visual system has to make guesses based on incomplete information. It is a fact that if an object exists then a drawing of it will follow the laws of perspective. However, our visual system follows an *incorrect* rule: that if a drawing follows the laws of perspective then a corresponding object exists, or could exist.

It is worth mentioning that the issue of interpretation is one of the barriers to developing the ability to draw faces: untrained people draw their interpretation and not what they see, with the result that they can draw a face more accurately if it is presented upside down.

Vivid evidence of the brain's construction of images is provided by autostereograms, one of which is shown in figure 1.5. At first sight a random collection of dots, if you focus on a point behind the image, after a period of up to a minute a three-dimensional picture of an oval with a square hole should emerge.[6] The effect depends upon the fact that we have two eyes, which can be persuaded to look at different parts of the autostereogram. The following experiment is well worth trying. Get a small piece of card and hold it close to your face and slightly to the side of one eye while you look at the autostereogram. Now move the card slowly until it partly covers one pupil. The result is that the part of the picture which is only seen by one eye returns to its random appearance while the part still visible to both eyes retains the three-dimensional image. Nevertheless both parts are equally clear. This is a particularly effective way of isolating the part of the visual system which constructs three-dimensional effects. The *information* needed to construct the three-dimensional picture is of course in the autostereogram, but the picture itself is not.

When we look at the world our brains decide that some objects are stationary in spite of the fact that as we look at different parts of them, the image on our retina is continually changing. Unless we are almost asleep, our brain factors such changes out before the mental image reaches our consciousness, informing us only of its current conclusions about which of the objects seen are stationary and which moving. Our 'mental world' is quite distinct from the constantly moving image on our retinas. The compensation mechanism is very specific and fails if one closes or covers one eye and presses the other eyelid *gently* from the side. Presumably the reason for this is that there has been no need for our brains to take into account the possibility of visual changes caused by pressing eyelids!

The above are a tiny fraction of the interesting ideas in this rapidly developing field. We have not listed the thirty-five specific rules of visual interpretation which Hoffman describes. These control what we think we see, which may or may not be correct in particular circumstances. We should not be surprised about this: natural selection worked to ensure that in the kind of circumstances

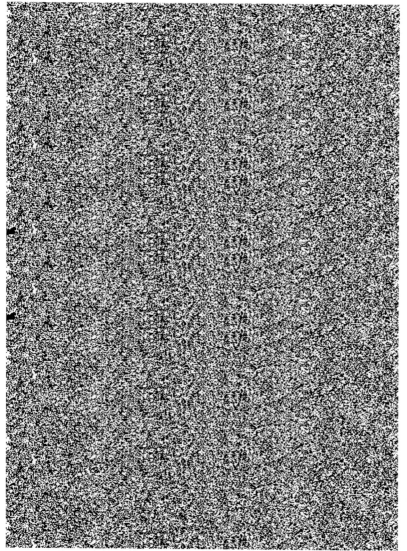

Fig. 1.5 Oval with Square Hole

Drawn using Randot v1.1 software written by Geoffrey Hausheer

we evolved in, our responses to visual stimuli normally promote our survival. It did not work to ensure that in the very specific situations dreamed up by psychologists the interpretations should bear any relationship with the truth.

The fact that we recognize something as an illusion created by fallible brain machinery does not enable us to banish the mistaken impression. Of course, given our intellect, we can often compensate for the mistake in a way which other animals almost surely cannot. But Hoffman points out that the situation is worse than this. Certain aspects of our visual interpretation are so deep-seated that we can hardly conceive that our mental constructions of the objects are quite distinct from the objects as they really are. It is necessarily difficult to expand on this idea, but he describes the analogy of computer games with multiple human players. The people involved have the feeling that they are carrying out actions in a virtual landscape, and it is clear that the interactions between the players have an objective aspect: different players agree about the progress and outcome of the game. On the other hand, what is actually happening can only be explained in terms of a collection of electrical currents flowing through circuits inside several computers. So the mental experience is caused by an artificial system whose nature is entirely unperceived by the participant.

It could be argued that the fact that we have rules of interpretation and that we may be led into error in some contrived situations has no philosophical importance: in all normal situations, if we have a subjective impression of a table in front of us, that is because there is a table in front of us, and this is what constitutes seeing the table. On the other hand, one does not need to be a philosopher to appreciate that we are only aware of the surface appearance of the table, and occasionally of its weight. The manufacturers of rosewood tables exploit this by restricting the rosewood to a thin surface veneer. If our sense organs enabled us to 'see' the interior of tables, this cost-saving device would fail utterly.

There are quite ordinary situations in which what we see has an obviously uneasy relationship with what is there, the most obvious being when we look at a mirror. The image we see *seems* to be behind the glass, but we interpret it as being a reflection. Our ability to make this interpretation is shared by very few other mammals, even though their eyes have similar structures to our own. Even we occasionally find it hard to relate to this image: when I was younger I made frequent efforts to cut my hair in a mirror, but never really mastered the skill. A person who could not recognize himself in a mirror would be abnormal by our standards, but would be no worse off than most animals. We, in turn, would be regarded as grossly mentally deficient by an alien which could cut its hair in a mirror without effort, which recognized faces upside down as easily as if they were the right way up, or which could 'see' the route through a complicated maze drawn on paper without conscious effort. What seems straightforward and obvious is, in fact, highly species-dependent. It depends entirely upon what unconscious processes your brain is capable of carrying out.

It might nevertheless be said that in the above case one sees an image 'of oneself' in the mirror, and it is related in an objective fashion to how one actually is, so one does indeed see oneself. Now imagine a future world in which all public advertisements make use of holograms, and in which television newscasters are computer simulations of people who are long dead or never existed. Using technology which almost exists today we may be surrounded by images which are not based upon any real object. We would be seeing something, but it would not be what it seems to be, nor would anyone think that there should exist any objects relating to what they see in everyday life.

Some of the above examples might seem to be frivolous, but when we get to the discussion of quantum theory we will confront the possibility that our brains may not be capable of constructing *any* comprehensive visual model of what is going on. The quantum world is really and objectively there, but it is so remote from the world in which we have evolved that we may never be able to construct an intuitive model of it. Almost every physicist agrees that the real nature of quantum particles remains beyond our imagination, and most probably agree that the only comprehensive model we will ever have of quantum theory will be a purely mathematical one.

Disorders of the Brain

The last section concentrated on the normal properties of the visual system, but there are many perceptual abnormalities (agnosias) which result from damage to particular parts of the brain. These further demonstrate the extent to which our view of the world depends upon interpretation within the brain. One of these, called cinematic vision, occurs when a person with perfectly clear eyesight is unable to recognize motion. The person afflicted sees a series of still views of objects, so that a car approaching is seen first as a small vehicle in the distance and then suddenly as a much larger one close up. Similarly a sufferer trying to pour a cup of tea may first see a static tube joining the teapot to the cup, and then suddenly a large pool of tea covering the table.

In blindsight a person is not consciously aware of objects in a certain part of the field of vision, even though their eyes are perfectly normal. When asked to guess what is present, and where it is, they are frequently correct, to their own surprise. Very recently brain scanners have provided evidence that images on the 'blind' side of the field of vision are processed differently from those on the normal side; the method of processing presumably bypasses whatever brings the perceptions to the consciousness of the person. These fascinating discoveries have the potential of providing deep new insights into the nature of the 'consciousness mechanism' in the brain, and are the subject of active research.

The term recognition agnosia refers to the inability to recognize an object by sight even when it can be recognized easily by touch, or the inability to recognize the faces of close friends and family even though their voices evoke normal responses, or the inability/refusal to recognize that one side of the body actually belongs to the sufferer. In 1985 Oliver Sacks described a patient who was a talented musician and able to engage normally in conversations in spite of the fact that he was unable to recognize most common objects visually. Particularly strange was that he seemed to accept his failure to recognize, say, a rose or a glove visually as entirely unremarkable, when he could give accurate descriptions of their parts and colours. This kind of mental loss is much more disruptive of normal life than would be the loss of vision, since it involves the partial disintegration of the personality.

Hirstein and Ramachandran have recently made an in depth study of a man who developed Capgras syndrome following a head injury.[7] Tests showed that he had no obvious deficits in higher functions and no evidence of dementia, in spite of the fact that he believed that close family members were impostors who looked exactly like the genuine people. Indeed he suffered the same problem with respect to himself. He recognized mirror images as being of himself, but would refer to photographs as being of another person who looked exactly similar; he sometimes even referred to himself as not being the genuine person. The best explanation of this syndrome at present is that there are two separate circuits involved in relating to close relatives, one dealing with recognition and the other creating an appropriate emotional response. If the circuit producing the emotional response does not function then it may be impossible for the unfortunate person to believe that the relative is who they seem to be. The fact that this may even apply to the person's response to himself raises a deep philosophical question about the nature of our self-consciousness. It appears that even this is not a unitary entity, but involves the correct interaction of a variety of independent modules. A provocative way of putting it is that our sense of self is created by the modules in our brains in order to help it to function.

Turning to mathematics, it has become clear that the ability to distinguish between very small numbers, those below about 4 or 5, does not involve counting but depends upon a specific module, probably in the left inferior parietal lobe. In *The Mathematical Brain* Brian Butterworth emphasizes that reasoning about even very small numbers involves a specific mechanism. People whose number module is damaged, either from birth or because of a stroke, *may have perfectly normal intelligence* apart from the fact that they have serious deficiencies in any situations which involve even very small numbers. Some cannot see without counting that a group of three objects is bigger than a group of two similar objects. By timing how long they take to do simple comparison tasks, it has been discovered that they may find it as hard to distinguish between the pair 9, 2 as between the pair 9, 8. Such people either cannot cope with numbers bigger than 5 by formal counting, because they do not understand what counting *means* in its application to the real world, or they count very slowly and painfully and only up to rather small numbers. This problem is now a recognized disability called dyscalculia, and is sometimes associated with dyslexia.

We like to believe that many of the skills mentioned in the paragraphs above are matters of general intelligence, but they are not, and this must undermine the classical view of consciousness and rationality as unified entities. If we act rationally in some situation this is because each of the modules in our brain behaves appropriately *in that situation*. This being so, one can imagine our distant descendants possessing extra modules in their brains whose function we are incapable of comprehending, and which enable them to understand matters entirely beyond our mental grasp. Our invention of language and subsequently of science have enabled us to progress far beyond what our unaided minds can grasp, and have led us into territories which we could never have entered without their support, but there are still limits to our mental capacities. Some indications of the extent of these will be presented in later chapters.

The World of a Bat

It is well known that the vision of frogs is dramatically different from ours. They do not see static objects, and can only react to motion. As a result, if surrounded by recently killed insects they will starve, but as soon as one flies across their field of vision they can react appropriately. In this section, however, we will discuss bats, because their quite different type of perception cannot be so easily dismissed as just an inferior version of our own.

When we consider the perception of bats below, we will be referring to their echo location system, and not their vision. Because bats emit high pitched series of clicks and are aware of the time delay and pitch of the echoes, they have precise information about the distance and rate of movement of obstacles or prey. This has some quite important implications for their perception of the world. The first is that distant objects must appear much darker, or dimmer, to them than closer objects, because the intensity of the echo from an object decreases very rapidly with its distance. For humans the apparent brightness of an object stays the same as it moves away, and only its size decreases. More importantly it is likely that a (hypothetical intelligent) bat would not consider that a picture of an object has any similarity to the object itself. Since its radar builds in distance information, a picture must appear to a bat to be a flat pigmented rectangle, quite unlike the three-dimensional object which it seems to resemble in our minds. We appreciate flat pictures because our vision is essentially two-dimensional, but the bat would be correct in maintaining that there is no physical similarity.

We cannot really know what subjective impressions bats experience, but the following thought experiment may help. Let us try to imagine what vision would be like in a world in which green light moved through the air much more slowly than red light. When viewing a static object the time delay for the arrival of green light compared with red would make no difference to our perceptions. Now suppose the object starts to move to the right. The red image emanating from the object at any moment reaches our eyes slightly earlier than the green image produced at the same moment—correspondingly, at any moment we see a red image which was produced at a slightly later time than that at which

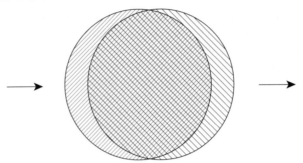

Fig. 1.6 Colour Fringes

the green image was produced. The result is that the object acquires a red fringe on its right side and a green fringe on its left side, as in figure 1.6—in which the colour fringes are replaced by hatching. Since we could already see how fast the object was moving, we could use this effect to draw conclusions about its distance: the further away it was the thicker the fringes would be. Let us now imagine that a module in our brains could interpret the colour fringes *before* they reached the conscious mind. Then we might have an enhanced three-dimensional perception of objects, but only if they were moving across the field of view. Finally suppose that the object is moving straight towards us. Then its boundaries are expanding on all sides, so it will be completely surrounded by a red fringe, and once again we might be able to perceive its rate of approach to us particularly clearly while it remains moving. These extra senses would be extremely valuable in a society which is so heavily dependent on cars.

Suppose instead that the S (blue) colour receptors in our retina responded not to the colour of light but to the distance of the object being viewed, while everything else about our colour vision was unchanged. Then we would look around and see objects with various shades and combinations of colours as at present. However, we would know that the more blue an object was the closer it was. This would provide a much enhanced sense of depth. It might be possible to implement this idea using modern computer processing and virtual reality displays, and it might even be useful to people such as pilots of aircraft. Perhaps this idea has already been patented!

What Do We See?

In the early days of research on vision, it was believed that the image falling on the retina was mapped with some modifications onto a part of the brain, where our mind became conscious of it. This led to the joke about a homunculus inside the brain 'looking' at the image laid out somewhere there. As a result of years of experimentation we now have a very different picture. The image falling on the retina is torn into fragments, so that edges,

colours, motion, and particular shapes such as mouths and eyes are all analysed separately. There are even specialized neural circuits which detect only
edges with particular orientations. At the end of this process a new and quite
different 'image' is constructed, which we commonly suppose to be a 'true'
representation of the original three-dimensional object. If any of the separate
modules which process different aspects of the original image is damaged by
a stroke, or functions incorrectly because the image is highly unnatural, then
we get at best an optical illusion and at worst a completely incomprehensible
result.

According to experimental psychologists, our subjective impression is never
of the object as it is. It is a construction which enables us to behave appropriately
in almost all ordinary circumstances. Evolution has ensured that our constructions give us a *useful* picture of reality, one which generally helps us to survive.
These experimental findings should encourage us to re-examine the way in
which we relate to our everyday surroundings. People rarely think about the
extent to which we are obsessed by the surfaces of objects. Objects are three-
dimensional and most of their material is inside them, not on their surface. How
many of us ever think in a tactile as opposed to an intellectual manner about
the thousands of kilometres of ground underneath us? The existence of these
things is known rationally, but our senses do not inform us about them, so we
ignore them. Presumably cows have no concept that there might be anything
underneath the earth and grass they stand on, even though their vision is quite
similar to our own. On the other hand those of us who live in the countryside
often contemplate the stars in the night sky, which are far more remote, simply
because our senses do inform us about their existence.

To the extent that we have a correct or true view of reality it is *a result of the use of our intellects* rather than simply because of the evidence of
our senses. Over many centuries we have learned that the Sun is stationary
although it seems to move, and the Earth rotates although it appears to be stationary. We have learned that a table is almost entirely composed of discrete
atomic nuclei and electrons separated by empty space, although it appears to
be solid and continuous. We have come to accept that TV programmes can
travel through empty space to our sets even though our senses provide no direct
evidence of this. We devote enormous technological resources to the avoidance
of infections by invisible particles called bacteria and viruses. These facts, and
many others, show how heavily our interpretation of reality depends upon the
technical knowledge accumulated by the society we are born into.

The idea that our instruments provide a truer picture of reality than our
senses arose in the seventeenth century. It was a key ingredient in the scientific
revolution, to be discussed in Chapter 6. Robert Hooke expressed it as follows
in *Micrographia*, published in 1665:

> *The next care to be taken, in respect of the Senses, is a supplying of their
> infirmities with Instruments, and, as it were, the adding of artificial Organs
> to the natural; this in one of them has been of late years accomplisht with
> prodigious benefit to all sorts of useful knowledge, by the invention of Optical*

> *Glasses ... It seems not improbable, but that by these helps the subtilty of the composition of Bodies, the structure of their parts, the various texture of their matter, the instruments and manner of their inward motions, and all the other possible appearances of things, may come to be more fully discovered; all which the ancient Peripateticks were content to comprehend in two general and (unless further explain'd) useless words of 'Matter' and 'Form'.*

The main change since Hooke wrote those words is that he thought primarily in terms of augmentations of existing senses, whereas modern instruments provide us with 'senses' quite unlike any which we naturally possess.

Our new reliance upon instruments is not as straightforward as it appears. They do not tell us anything about reality until we interpret the readings we obtain from them in the light of some theory of how they work. We are convinced that this is not a circular process by the huge variety of independent sources of confirmation of the picture which we have built up over centuries of scientific investigation. I shall have more to say about this in Chapter 10.

1.3 Language

Physiological Aspects of Language

The visual system of humans is amazingly sophisticated, but it is not radically different from that of other mammals. Many experts consider that the best bet is that our specifically human intelligence is related to our use of language.

Although language is clearly very important, it is easy to be carried away by this line of argument. In a different context the philosopher Bryan Magee has argued persuasively that many of our high level judgements and skills have no verbal component at all.[8] Playing a violin, discriminating between wines, judging whether someone is trustworthy, admiring a painting, deciding whether two colours clash—all of these activities can occupy our full attention without being in any way verbal. Magee writes that even when one is struggling to write down one's thoughts, one has to know what one wants to say *before* one chooses the words which express it best. Writers frequently revise sentences again and again, a nonsensical situation if one believes that their deepest thoughts are already verbal in form. Clearly there is more to being human than possessing language, but language has the advantage among our skills of being easy to investigate. With due apologies, I will therefore concentrate on what is known about it, while hoping that eventually scientists will move on to the consideration of our other peculiar skills.

It is well known that the structure of adult human throats is substantially different from that of all other mammals, and that this enables us to produce a much wider range of sounds than, for example, chimpanzees. Like most mammals, human babies have a relatively high larynx which connects to the nasal cavity when swallowing, so that babies can breathe at the same time as suckling. The position of children's larynxes drops by the age of seven, and has the unfortunate consequence of making us uniquely susceptible to choking on food. This design fault results in a significant number of deaths every year and could not exist unless there was an important compensating advantage. It is clearly a genetic adaptation enabling us to communicate by speech more efficiently. It would be strange if the changes in our vocal apparatus were our only adaptations to the use of language, and there is in fact plenty of evidence for the existence of a specific inborn language ability. There is an inherited disease, called Specific Language Impairment, which does not involve impairment, of the general intelligence. Conversely people with Williams' syndrome, associated with a defect on chromosome 11, are very fluent conversationalists with large vocabularies, but their IQ is typically around 50.

There have been a few well documented cases of children who have not had the opportunity to start learning to speak until an advanced age. If they start before the age of about six, they are usually able to catch up the missing ground and develop normal speech skills. If they start learning to speak after that age the task becomes steadily harder and the eventual skill acquired becomes progressively lower. More compelling, because of the numbers involved, are surveys of the acquisition of English by Korean and Chinese children who have immigrated into the USA at various ages. If they arrive by the age of six, then their eventual language skills are indistinguishable from those of people born in the USA; for those who arrive at a later age the eventual ability in speaking English depends upon the age of arrival.[9]

This is related to the existence of critical periods for the acquisition of a number of skills, and is explained in terms of neural systems degenerating or being rewired if they are not used at the 'expected' stage of development. Thus kittens brought up in a limited environment with no vertical lines are later unable to distinguish them: the relevant part of their visual cortex is redeployed if it is not stimulated during the critical period. The ability of human adults to discriminate sounds is strongly dependent upon their own language: many Europeans simply cannot hear the differences between different Chinese names because they are not sufficiently sensitive to pitch. Similarly the difficulty which Japanese have in distinguishing between the sounds 'l' and 'r' is based upon changes in the physical circuits in their brains; this occurs in response to the range of sounds they hear around them from a very early age.

Another type of evidence for specific language skills is the astonishing rate at which words are learned in the early years. Tests of USA high school graduates show that on average they know the meaning of about 45,000 words, or up to 60,000 if one includes proper names. This implies that they have learned about nine words per day since birth, most of which are acquired without any apparent effort. Indeed in the first few years of life the rate of learning is even higher. Contrast this with children's difficulties in learning to read. Here progress depends upon formal education programmes, which require considerable perseverance on the part of both teachers and children. Although almost nobody fails to learn to talk, the number of people who are illiterate is very substantial, in most cases because of inadequate teaching. The implication is that we have evolved neural circuits which make spoken language easy to acquire, but that this has not happened for writing in the six thousand years since the first written language appeared.

The language instinct of humans is genetic in origin. This does not mean that there are genes for particular grammatical constructions, nor does it imply that there must be a deep 'Universal Grammar' as Noam Chomsky has argued. Genes code for the production of proteins, and the route from proteins to specific language skills is bound to be complicated and indirect. The fact that a 'faulty' gene may lead to a particular failure in grammar production does not imply that the gene is responsible for that feature: the failure of a resistor may stop a TV working, but that does not mean that the resistor is more responsible for the picture than several hundred other components.

There is much evidence that the use of language enables us to memorize events much more precisely, because the stimulation associated with the use of language facilitates a further spurt of brain development. There have been extended attempts to teach chimpanzees the use of language by bringing them up in human family environments. Since they do not have the vocal apparatus for speech, they have been taught using American sign language. It has proved possible to teach chimpanzees up to a few hundred words in their first five years of life, a tiny fraction of what human children achieve.[10] The comparative abilities of human children and chimpanzees are rather similar until the point at which language develops in the children, somewhere between their first and second birthdays, after which our mental development accelerates away from that of chimpanzees. A related point is that we have very few memories of the period before we learn the use of language. It is obvious that our use of language does not merely enable us to communicate, but that it also profoundly affects the way we perceive the outside world.

Recent experimental evidence confirms that the environment in which animals live changes the physiology of their brains. Post-mortem examinations show that rats raised in an enriched environment have thicker cerebral cortexes with more nerve fibres than other rats. Until recently it was thought that brain structure is largely fixed by adulthood, but there is now evidence that when middle-aged rats are placed in an such an environment, their brains grow substantially. The following two examples provide recent evidence for the effects of learning on the wiring of neurons in *human adults*. It appears from a variety of recent experiments on both humans and monkeys that certain types of repetitive strain injury suffered by typists and musicians are *not* caused by damage to the tendons. It appears that the abnormal use of the affected digits eventually leads to the brain rewiring the relevant circuits in a manner which prevents them working properly. The abnormal neural connections have been observed directly, and in some cases appropriate retraining can reverse the problem by causing the brain to re-wire the neurons back to a more normal pattern.

London taxi drivers are required to pass very demanding examinations relating to street layout and navigation: acquiring the necessary skills may take a few years. It has recently been discovered that their right posterior hippocampuses enlarge slightly but progressively the longer they do their jobs. The fact that this change is progressive demonstrates that it is related to the actual acquisition of the spatial skills. It provides good evidence that the brain retains substantial plasticity into adulthood.[11]

The extra stimulation we receive from the use of language almost certainly leads to the formation of extra synaptic connections in early childhood. This in effect makes us into a different animal from what we would be in the absence of such stimuli. We can easily imagine a feedback cycle operating between the development of society and of the human brain. When the adults of a tribe develop skills which aid its survival, their young learn those skills more rapidly because of their greater brain plasticity, and this makes it easier for them to develop new skills of a similar type. The size of this effect would depend on the degree to which the structure of the brain is set at birth. We know that for primates and particularly humans this is very low by comparison with other animals. As pre-human and prehistoric societies became more complex the most successful individuals might be the ones whose brains were the *least* fixed at birth, because they would be the most able to learn the skills which their culture required. They would survive to breed more frequently and pass on their superior ability to learn to their offspring. This would enable another round of development of the complexity of the social group. Eventually, this process leads to a genetic change in the species by purely Darwinian mechanisms.

The above already shows dangers of developing into a 'Just So' story, and we will not pursue it further. Many hypotheses have been put forward concerning the reasons for the initial development of language, but it is difficult to test them scientifically. One idea depicts human intellectual development as the progressive growth of ever more sophisticated strategies for the purpose of deceiving and gaining advantages over neighbours. Even the date at which the human throat developed in its present form is unknown, because it is composed of soft tissues which are not preserved after death. We do know that sophisticated stone technology and cave painting existed about forty thousand years ago, when homo sapiens was already well established, but much of what is written in this field has a rather slender factual basis.

What are the implication of these ideas for our mathematical abilities? It is probable that we did not need the ability to count to more than a dozen or so until the last ten thousand years. Current research indicates that the ability to distinguish numbers up to about 4 depends upon circuits which act at a pre-conscious level.[12] It appears that formal computational arithmetic uses circuits in the brain which are also involved in generating associations between words. In contrast numerical estimation shows no dependence on language and relies primarily on visio-spatial networks of the left and right parietal lobes. Together these results suggest that human estimation skills, which are shared with animals and already present in pre-verbal infants, have a long evolutionary history. On the other hand our development of advanced mathematics could only have arisen within a culture possessing a formal system of education.

It is evident that we do not have sense organs which enable us to perceive the meaning of large numbers such as 127928006054345 or to gaze directly at some abstract mathematical world. Our reason is not a kind of a sense organ: the knowledge which we obtain using it depends heavily upon the culture and period in which we live. People become good at mathematics for the same reason that they become good at swimming or music, by devoting their energies to developing the relevant skills over a sufficiently long period. They become truly outstanding by being obsessively interested over a period of ten years or longer.

Ramanujan was just such a person. One of the most extraordinary mathematical geniuses of the twentieth century, he was born in India in 1887. As a child he displayed ability in a wide variety of subjects, but from the age of fifteen started to devote himself to mathematics to the exclusion of all other interests. By conventional standards his knowledge was extremely limited, but he developed insights into number theory which led Hardy to invite him to Cambridge, England in 1914. Before his death in 1920 from a protracted illness he had written down enough unproved new results to keep other mathematicians in work for several decades. His best parallel in more recent times may be Paul Erdos, a Hungarian who literally lived for mathematics, abandoning any semblance of a normal life as he wandered from country to country seeking problems to test his wits on.

Of course mathematicians are not merely people who are good at arithmetic. There is little likelihood that we could have evolved any specifically mathematical genes over the last few thousand years, but the following facts hint at one of the possible sources of mathematical ability. Many mathematicians have particularly strong spatial imaginations, in common with architects, artists, and brain surgeons, and this might well have had advantages for hunter-gatherers travelling large distances every year. Spatial ability seems to be somewhat correlated with left-handedness, which is in other ways (increased susceptibility to auto-immune diseases and decreased life span) an evolutionary disadvantage. Left-handedness is also partially inherited and may be an example of a balanced polymorphism.[13] Mathematical ability may be a result of combining the functions of the basic number module, spatial visualization skills, and general reasoning powers, reinforced by appropriate education from an early age. The extent of this ability is perhaps surprising, just as the development of a trunk in the elephant is surprising; but ultimately there are no deep philosophical conclusions to be drawn from the ability of a very small proportion of people to do advanced mathematics. It is a contingent reality. If it were not so we would no doubt devote our considerable energies to puzzling over some other issue.

Social Aspects of Language

Vision provides information about the immediate environment, but the great majority of speech involves remote events or social interactions. The purpose of this section is to demonstrate that the understanding of even simple sentences involves enormous prior knowledge. It is also relevant to arguments against scientific reductionism, discussed in more detail in Chapter 9. Consider the following sentence, entirely typical of those which make up our everyday conversation.

> *Joanna is happy because her daughter, Catherine, has done well in her A level examinations.*

The implication that Joanna is the mother of Catherine is not as straightforward as it appears. Society has already divided motherhood into three different types, legal, genetic, and womb motherhood, and it is already possible for a child to have a different mother of each type. It may soon be possible for a child to have a womb mother and a clone father, but no genetic mother. What will have become of the concept of motherhood in another hundred years is anyone's guess. What is certain is that Aldous Huxley's *Brave New World* can no longer be regarded as science fantasy.

From the two names and the reference to A levels we may reasonably guess that Joanna is British. This is not the same as saying that she has a British passport, since the passport might have been obtained by a bribe. Nor does it mean that her ancestors were British. The peculiar nature of this concept is illustrated by a shameful episode in 1968, when the British Government introduced the concept of patriality in order to reduce the number of East African Asians who could enter the UK using their British passports. Effectively the Government decided tō split the concept of British nationality into two for political reasons.

The concept of examination is very important in our society, but it is indeed a concept, not a physical event. British schoolchildren prepare for examinations by undergoing mock versions, which are done under more or less identical physical conditions to the true examinations. The main difference between the true and mock examinations lies in the beliefs of the pupils and others about their significance.

We have seen that the simplest of sentences can combine concepts of a very abstract character. A few of these are objectively physical, but *most* refer to complex social institutions. Let us now look at the sentence as a whole. This might have related to a real occasion or be from a novel, but because of the context of this book you read it in quite a different way: the issue of concern was the interpretation of everyday sentences. We conclude that the significance of a sentence may be entirely altered by the context in which it was written. In fact many people believe that language evolved to facilitate social interactions rather than to communicate information about the outside world.

There is good support for this in today's world. One of the reasons why (British and probably all other) politicians are so annoying is that we, their audience, keep hoping that they might answer the questions which they are asked. They are playing a completely different game, namely using the interview or speech to persuade people to vote for them. They have achieved positions of power precisely because of their ability to deflect difficult questions, and to turn people's attention to issues which will show them in a better light. Scientists (and many others) tend to think that questions should be answered honestly, and languish in obscurity because we do not have the skill to use words to such advantage.

Objects, Concepts, and Existence

Although much of our daily use of language is heavily linked to our social structures, we also use it to analyse the world around us. The goal of this section is to establish that language frequently does not truly reflect reality; this problem is not capable of resolution because the number of words we can remember is far more limited than the variety of phenomena we wish to describe. For example, it is obvious that colours merge into each other continuously: there is no point in the passage from red to yellow through various shades of orange at which one can point to a physical or psychological boundary. Nevertheless we use discrete colour words. While the number of these can be increased, the boundaries between them are bound to remain artificial.

Consider next an example much loved by philosophers 'no bachelor is married', relying on the dictionary definition of a bachelor as an unmarried man. On logical grounds this seems impeccable. The problem is that, in English at least, dictionaries do *not* define the meanings of words: they only summarize how they are used in the real world.[14] This use changes over time. Thus none of the following accords well with the normal use of the word bachelor, in spite of the dictionary: a man living with a long term partner but not married to her; a recently widowed man; a forty year-old man who has been in prison since the age of fifteen. On the other hand a man who is permanently separated from his wife might well be called a bachelor. The phrase 'bachelor girl' suggests that at present the word bachelor has more to do with a life-style than with being male and unmarried. Of course this may change again. The Oxford English Dictionary has caught up with the fact that *independence* is now a key requirement in its definition of bachelor girls, but not for bachelor men. Even the nuances involved in the use of the word 'girl' are fraught with difficulties: university staff need to be careful about using it when referring to women students, even though only a small proportion would be offended.

With the dangers of over-simplification in mind, we now turn to the word 'existence'. Issues related to this word are at the core of many of the problems of philosophy.[15] The most elementary type of existence is that of material objects, such as the Eiffel Tower. Many other objects are not accessible to us, simply because of the passage of time, and their past existence has to be inferred from documentary evidence. Going beyond that, I believe that my ancestor one thousand generations ago in the female line had a navel, even though I have no direct evidence for the existence of the ancestor, let alone of her navel. In this case my belief is based upon the acceptance of certain regularities in nature;

this belief is not shared by those who consider that the world was created in 4004 BC.

Existence problems are closely related to questions about truth. As soon as one admits even the remotest possibility that some everyday fact might be false, one is admitting that one does not *know* that it is true, but only believes that it is so. Possibly we, as finite beings, have no access to final knowledge, and have to content ourselves with the statement that certain statements have such overwhelming evidence in their support that it makes no sense to regard belief in them as provisional. There have been endless debates about the relationship between truth, belief, and evidence which we must pass over here. Let me only add that one is already taking a realist philosophical position by assuming that beliefs about the past are either true or false.

I must confess to finding abstract discussions of such problems unrewarding, and prefer to consider particular examples which illustrate the difficulties which any general theory has to face. So let us discuss the nature of black holes, studied by Stephen Hawking and Roger Penrose between 1965–70. Their development of the earlier, non-relativistic theory of black holes depended upon the general theory of relativity, and led to the following conclusions. If a star is sufficiently massive (a few times the mass of the Sun) then it eventually turns into a supernova. The remnant after the supernova explosion may still be so massive that any light or other radiation which it emits is unable to escape beyond what is called its event horizon. In the theory the remnant, called a black hole, is invisible, but it may still have gravitational effects on other nearby bodies. There is steadily increasing evidence, many would say virtual certainty, that such objects do exist. A well documented example, Cygnus X-1, is the invisible component of a binary X-ray system in the constellation Cygnus; among many other candidates are V404 Cygni and Nova Scorpii 1994. In the last few years astronomers have found exciting evidence that most and perhaps all galaxies have supermassive black holes at their centres. The one at the centre of our own galaxy has just been identified as Sagittarius A*.

In spite of the accumulating evidence confirming theoretical predictions about the properties of black holes, the physics of the interior of black holes is not understood. General relativity tells us that there are singularities at their centres, but the physics of space-time near the singularities may only be explicable using quantum theory. If we believe in general relativity, we can never obtain any direct or indirect evidence about what is happening inside them. So we are expected to believe in the existence of something which is in principle unknowable—almost a religious injunction, except that it is made by obviously serious scientists.[16]

Although rainbows look like material objects, a little reflection shows that this cannot be true. Different people standing a few metres apart might agree that they are looking at the same rainbow, but disagree about where it meets the ground. Someone who appears (to someone else) to be standing 'in' a rainbow would not experience any peculiar visual effects. The simple fact is that rainbows are neither material objects nor concepts: the raindrops which cause our 'rainbow sensation' do not have any intrinsic properties to distinguish them from other neighbouring raindrops. They are only distinguished in terms of their spatial relationship with both the sun and ourselves as observers. Physics explains the phenomenon perfectly, but the structure of our language does not provide an obvious category into which they fall.

We turn next to concepts. Jerry Fodor[17] suggests that a concept should be defined as a list of 'features' stored in memory, that specifies relevant properties of the things the concept applies to. Fodor's definition does not necessitate the existence of *any* things to which the concepts apply, and it places concepts firmly within the realms of space and time. Thus, in spite of the fact that we believe that unicorns do not and never did exist, we have a reasonably clear idea of what they would be like. The concept is associated with definite features, and if someone uses the word without respecting those features, they would be using it incorrectly. In heraldry, art and sculpture lions and unicorns have exactly the same status; the important issue is whether the concept is clear, not whether the animals exist.

I was very embarrassed many years ago to discover that there was a suburb of South-West London called Surbiton. It had been mentioned frequently in newspapers as representing a certain type of middle class suburban political attitude, and I had concluded from the over-appropriate name that it was an invention. When I discovered during a rather confusing conversation that it actually existed, I was interested to realize that I did not need to change any of my other beliefs about its characteristics! Much later I realized that my original attitude towards Surbiton had been closer to the truth than I had thought. Many people living there no doubt regarded its newspaper image as a caricature. Its existence was irrelevant: if there had been no Surbiton, newspaper columnists would have chosen some other place to represent these particular attitudes.

There is a category of entities which are neither physical nor mental, but which exist as a part of our collective culture. The philosopher Karl Popper has argued that one should accept that something such a Roman law exists, because it has observable effects on the world of physical objects: people might end up in prison because it exists when they would be executed if some other legal system existed. On the other hand it is also clear that Roman law is a human construct—five thousand years ago it did not exist. In this respect justice is rather a more difficult notion. Some would say that it could not predate human society and is a biologically innate concept, while others might believe that it emanates from God and has always existed. Another example of an entity which exists by social convention is money, to be discussed in Chapter 9 in an argument against scientific reductionism.

Yet another type of existence is that of skills, such as producing an axe by knapping a stone, or playing a piano. Their existence can be proved beyond challenge by the person showing that they can perform the relevant task. A person can prove that they *understand* the skill in conversation, but they can only prove that they *possess* it by demonstration. The philosopher John Lucas has suggested that much of mathematics depends on the development of such skills rather than on the existence of abstract theorems.[18]

The peculiar relationship between time and existence is provided by the game Eternity, in which 209 irregular shaped pieces are assembled in a jigsaw-like fashion on a specially designed board. In 1999 the inventor of the game, Christopher Monckton, offered a prize of a million pounds to the first person who put the entire jigsaw together correctly. The game became very popular and, to Monckton's great discomfort, the prize was claimed in September 2000 by two Cambridge mathematicians, Alex Selby and Oliver Riordan. So one can have no doubt that a solution exists: it has made two people much richer and another much poorer. On the other hand its solution presumably could not have existed before the game was invented, so its existence has to be regarded as time-dependent. If one believes that the solution came into existence as the game was invented, should one symmetrically believe that the solution will disappear if all memories of the game are one day lost to humanity? And if some historian comes across a description of the game in some library, does the solution then come back into existence immediately or only when someone rediscovers it? If it is the same solution, where was it in the intervening period? Are these real questions or are they just about how we choose to use the word 'exist'? We leave the question at this point, because the philosophical literature on such matters is vast, and does not appear to have led to a clear conclusion.

Numbers as Social Constructs

There are two extreme views about the nature of numbers, and many others in between. One, called mathematical Platonism, declares that numbers exist independently in some objective sense, and that mathematicians are engaged in uncovering the properties which they already have. The other declares that numbers are concepts of the same type as all others in our language, invented by us, and endowed with properties which we can then investigate or modify as we see fit. The issue is not about whether numbers exist, but whether they do so independently of society or as social constructions (concepts).

The Platonic position seems to be supported by the following argument, discussed at length by Benacerraf and others.[19] We are not permitted to use the word truth in mathematics differently from the way we use it in all other contexts. Therefore the statement that there are three primes between 45 and 60 must be true because it *refers* to entities which do indeed have the properties stated. We can examine these entities (the numbers between 45 and 60) one at a time, and confirm that exactly three of them are indeed primes.

As with all philosophical arguments, there are counter-arguments. Statements in ordinary language are extremely varied. Thus:

There are six types of outcome to a game of chess.

is a perfectly normal sentence, whose truth is certainly not based upon examining all possible games of chess and dividing them into exactly six groups according to their outcome. If there is any reference it is to the concept of an ending.

In other cases an apparently simple statement becomes steadily more obscure the longer one thinks about what exactly is being referred to. Consider the sentence:

There are five vowels in the English language.

The objects being referred to here are vowels. But what exactly are vowels? Since we consider that 'y' is sometimes a vowel and sometimes a consonant, it follows that being a vowel depends upon context rather than shape. It has some relationship with pronunciation, but in English one cannot decide what vowel is being used from the pronunciation. Every letter may appear in many fonts and sizes, so letters are certainly not copies of material objects. Once again we are referring to abstract objects, which have changed over time and even now vary from one language to another.

The above examples indicate that the possibility of making statements about abstract entities does not imply that those objects exist independently of society. According to the philosopher Karl Popper, numbers are also simply a social construction.

> *The infinite sequence of natural numbers, 0, 1, 2, 3, 4, 5, 6, and so on, is a human invention, a product of the human mind. As such it may be said not to be autonomous, but to depend on World 2 thought processes. But now take the even numbers, or the prime numbers. These are not invented by us, but discovered or found. We discover that the sequence of natural numbers consists of even numbers and odd numbers and, whatever we may think about it, no thought process can alter this fact of World 3. The sequence of natural numbers is a result of our learning to count—that is, it is an invention within the human language. But it has its unalterable inner laws or restrictions or regularities which are the unintended consequences of the man-made sequence of natural numbers; that is, the unintended consequences of some product of the human mind.* [20]

In *What is Mathematics, Really?* Reuben Hersh writes in similar terms, but with the advantage of actually knowing about mathematics from the inside.

> *Mathematics is human. It's part of and fits into human culture. Mathematical knowledge isn't infallible. Like science, mathematics can advance by making mistakes, correcting and recorrecting them. (This fallibilism is brilliantly argued in Lakatos's* Proofs and Refutations.*) . . . Mathematical objects are a distinct variety of social-historic objects. They're a special part of culture. Literature, religion and banking are also special parts of culture. Each is radically different from the others.*

Let me present an example which is relevant to the question of whether mathematical ideas and mathematical theorems are invented or discovered.[21] Perfect numbers are defined as numbers which are equal to the sum of their factors, including 1 but not including the numbers themselves. So for example

$$6 = 3 + 2 + 1$$

is perfect. The next three perfect numbers, 28, 496, and 8128, were all known to the Greeks, and a number of interesting theorems about them are known. In a similar spirit let us call a number *neat* if its number of factors, including 1 but not including the number, is a factor of the number. Thus 15 has the factors 1, 3, and 5, and the number of these factors, namely 3, divides 15, so 15 is neat. On the other hand 125 has three factors, 1, 5, and 25, and 3 is not a factor of 125 so 125 is not neat. It is easy to prove that every prime number is neat and that a product of five distinct primes is neat if and only if one of those primes is 31. A product of six distinct primes cannot be neat but a product of seven can. If p is a prime then the number p^n is neat if and only if n is a power of p. Other theorems about neat numbers could no doubt be proved if one were interested, and one could make conjectures about the asymptotic density of the neat numbers in the set of all numbers. Did the class (i.e. collection) of all neat numbers exist before I invented it, specifically in order to write this paragraph? I would contend not; mathematics could well do without the concept and it would probably never have been invented but for my wish to demonstrate how easy it is to produce definitions and theorems. On the other hand once I *invented* the class and then *discovered* the theorem about products of five distinct primes, its truth was not a matter of opinion. It can be tested experimentally on a computer, and proved using standard mathematical methods.[22]

This is entirely in accord with other contexts in which we use the word invent. When the Wright brothers invented the aeroplane, it was nevertheless an objective fact that it could fly. Nobody has invented a matter transporter of the type familiar to Star Trek fans because inventing something, as opposed to imagining it, necessitates that it works. Similarly in mathematics one cannot invent a concept if that concept is self-contradictory. An attempt to develop a theory of pentagons for which the sum of the internal angles is an odd number of degrees leads to only one theorem: no such pentagons exist. Mathematics is relatively objective in the sense that it does not often allow for varying opinions, but that by no means forces one to the conclusion that it must be about entities which pre-exist their first consideration by human beings. Whether or not a particular idea, be it primes or neat numbers, is ever invented, depends upon cultural issues and also on whether the idea is simple enough for our brains to understand. It may also be a matter of mere chance.

On the other hand declaring that numbers are 'merely' social constructions leaves some quite serious problems to be resolved. Few people would dispute that diplodocus had four legs, long before human beings evolved the ability to count them. Some people argue from this that numbers must have existed before we invented them. One may respond that what actually existed was the diplodocus with its various parts, some of which we choose to call legs in spite of the fact that the front ones are anatomically quite different from the rear ones.

Only after we have developed a sufficiently sophisticated language is it possible for us to formulate a sentence involving the number 'four'. We are then correct to say 'diplodocus had four legs' because this provides a good match between what we see and the concepts which we have constructed.

It has been put to me that if an alien civilization were found to have been counting before the human race evolved, that would prove that numbers exist independently of ourselves. While this is true, it is not terribly profound. If an alien civilization were found to have used diagrams (or bottles), this would prove that diagrams (resp. bottles) existed before we independently thought of them. The possibility that two totally independent civilizations might have some practices in common has no deep implications.

It is an interesting fact that although the use of diagrams has enabled us to organize our knowledge in a way not easily achievable otherwise, nobody appears to claim that they have some deep philosophical status. Yet diagrams dominate science almost as much as do numbers. In *The Origin of Species*, Charles Darwin did not use any mathematics but he did include a diagram, which he discussed for several pages; this was of a schematic tree showing the evolution of species from one or a few ancestors. The task of filling in the details of this tree has dominated evolutionary studies ever since. William Smith published the first geological map of Great Britain in 1815 after many years travelling and classifying the fossils which he found embedded in rocks. This map transformed geology into a true science and set the scene for all future work in the field. Mendeleyev's periodic table, which classifies the chemical elements into types, still appears on the walls of almost every university chemistry department. A more recent example is the use of flow charts to explain the interactions between parts of complex projects or organizations. All of these *can* be described using words alone, but at the cost of becoming more or less impossible to understand. We use diagrams because they present information in a manner which *our type of brain* can easily assimilate.

It has been suggested that the situation with numbers is quite different: it is claimed to be self-evident that alien civilizations *must necessarily* use numbers in the same way as we do, and this proves that numbers have an existence independent of any civilization. This is evidently pure conjecture. It is amazing that people are so confident that intelligent aliens will be essentially similar to ourselves, apart from superficial differences such as having two heads, tentacles, etc. On this planet we contemplate highly organized insect colonies and know that we will never be able to communicate with them. There have been arguments about whether dolphins have equivalent intelligence to our own, or even higher intelligence which we cannot recognize because of that very fact. How much less can we assert what undiscovered alien civilizations must be like, when we do not even have any evidence that any such civilizations exist?

Finally, it is claimed that the utility of mathematics in the understanding of the physical world is so striking that this proves that it cannot just be a social construction. Hersh answers this with the blunt assertion that our mathematical ideas in general match our world for the same reason that our lungs match Earth's atmosphere. One should add the caveat here that in the former case one must be referring to cultural evolution, whereas in the latter biological evolution was the driving force. But the point remains that most mathematics has grown from attempts to describe properties of the external world,[23] so it is not a coincidence that the two match. Indeed after more than two thousand years of development of the subject, it would be amazing if they did not.

Over the next chapters we will see evidence that mathematics is not quite as powerful as people would have us believe, and that some of its power only exists 'in principle'. In other words there are many phenomena (such as the weather) which no amount of mathematics will *in fact* predict in detail. Mathematics is indeed our best tool for understanding several branches of science, and it is extraordinarily good, but it will not enable us to resolve every problem we are interested in.

Notes and References

[1] This image of the nature of electricity is not scientifically accurate, but it enables us to relate its properties to things we are familiar with.

[2] Ronan 1978, p. 226

[3] Mollon 1992

[4] Mollon 1997, p. 390

[5] Davidoff et al. 1999, Shepard 1997

[6] Some people find it quite hard to achieve the effect the first time. Try removing your glasses if you wear any; experiment with defocussing your eyes, and putting your head at various distances from the page, which you *must* view sideways.

[7] Hirstein and Ramachandran 1997

[8] He was criticizing the Oxford school of linguistic philosophers.

[9] Pinker 1995, p. 291

[10] Some investigators have denied that the chimpanzees actually learn even this much, in spite of appearances.

[11] Maguire et al. 2000

[12] Butterworth 1999, Dehaene et al. 1998, Geary 1995

[13] This is an inherited characteristic which has both advantages and dis-advantages, preventing it from becoming either extinct or universal. Corballis 1991, p. 92–96

[14] The Académie Française has tried to regulate the French language much more strictly, but with decreasing success in recent years.

[15] I agree here with the Oxford philosopher Gilbert Ryle, who argued in *The Concept of Mind* that the word has more than one meaning and that the failure to recognize this was at the root of the Cartesian mind-body fallacy.

This book was written at the height of the Oxford passion for linguistic analysis, now largely spent. But there are indeed situations in which the vagueness of ordinary language can seriously mislead people.

[16] My colleague John Taylor has pointed out that we could indeed go into a black hole to find out if the predictions of some theory about them are correct, but we could not then tell anyone outside what our conclusions were. So a more correct statement is that we could never compare the insides of two different black holes.

[17] Fodor 1998

[18] Lucas 2000

[19] Benacerraf 1996

[20] Popper 1982, p. 120

[21] A similar discussion of prime numbers has been given by Yehuda Rav [1993], but it is better to avoid a topic which many people have already encountered.

[22] Postscript. It is amazing how events can overtake one. In 1998 Simon Colton's HR computer program invented almost the same concept [Colton 1999]. The only difference is that it counted *all* factors, not just proper factors, and so ended up with an entirely different class of 'refactorable' numbers. It then turned out that these numbers had already been discovered by Kennedy and Cooper without machine aid in 1990. So my neat numbers are still original!

[23] The most important exceptions are number theory and group theory. But group theory was co-opted into geometry long before its relevance to particle physics became apparent.

2
Theories of the Mind

2.1 Preamble

This chapter is largely devoted to discussions of the beliefs of Plato and Descartes. Why, you may ask, do we need to spend time discussing the views of two long dead philosophers? The answer is that their systems still exert a profound influence, in spite of their obvious faults. They have become so much a part of our culture that we rarely pause to examine them. Only by doing so have we any hope of breaking free of the constraints which they impose on our thinking.

What I have to say may appear negative, in the sense that I am pointing out major flaws in belief systems without proposing a detailed alternative. My defence is that it is better to acknowledge that we are not even close to an understanding of the true nature of the world, than to comfort oneself clinging to beliefs which stand no chance of being correct. Admitting this is hard, particularly for those who have devoted their lives to the search for final knowledge.

We start with a discussion of Plato, because many mathematicians declare themselves to be Platonists. For most this is just the simplest way of avoiding serious thought: they subscribe to almost none of Plato's beliefs and have worked out no neo-Platonist position. A few are more serious in their Platonism, and among these one must mention Roger Penrose and Kurt Gödel. I do not agree with anything they say (about this subject), but at least they are sufficiently honest to have formulated views with which one can argue. We will see some of the difficulties associated with their position.

We then turn to Descartes' argument that mind/soul and body/matter are entirely different types of entity. Although this has had an enormous influence on the development of science, nobody in the seventeenth century could explain how two entirely different types of entity could interact with each other, and subsequent philosophers have done no better. Of course many explanations were proposed, including the idea that God has arranged that thoughts and bodily actions would be synchronized, although there was no causal relationship between them. But nobody has devised an explanation which commands general assent. The successes of Western science have all concerned the behaviour of matter, and some scientists and philosophers believe that the independent existence of the mind/soul is an illusion. On the other hand the general population remain committed to mind-body dualism, which fits in well with religious belief provided one does not examine it too carefully. We consider a number of examples which show how confused the various current views are, and demonstrate how badly a new post-Cartesian approach to these problems is needed.

In the final section of the chapter we will turn to the current debate about the problem of the existence of consciousness. I explain *why* current computers should not be regarded as conscious, and that we ourselves are conscious of only a small proportion of the activity in our brains. The fact that some of the deepest forms of processing are not conscious suggests that our thinking is not ultimately fully rational or under our control. The precise mechanisms which correspond to conscious experiences may well be found within the next few decades, but this does not necessarily mean that we will ever be able to duplicate consciousness in machines.

2.2 Mind-Body Dualism

Plato

The influence of Plato as the founder of systematic philosophy has been immense, in spite of the fact that many of his arguments have been disputed or even rejected since his time. When discussing his writings, we face the problem that his views developed and even changed during his life. In the late work *Parmenides* he criticizes his own theory of Forms in a dialogue between Parmenides and Socrates, and it is often not clear what the conclusion of the discussion is. He even uses arguments which appear to be incompatible in a single book. The account below is therefore a selection from his views, and almost any of the statements made could be the subject of prolonged debate.

Plato frequently put words into the mouths of Socrates and others, and we often cannot tell to what extent these represented his own ideas or beliefs. The real Socrates lived in Athens in the fifth century BC, and was considerably older than Plato. He wrote nothing of his own, and is mentioned by several other Greek writers, but Plato is the main source of information about him. He was a considerable public figure, who was eventually condemned to death in 399 BC, ostensibly for 'corruption of the young' and 'neglect of the gods'. The actual reason was his close association with Critias and Alcibiades, who were on the 'wrong' side in the political ferment of that period. Plato's story of Socrates' refusal to accept a lesser punishment, or to attempt to escape, is probably true. He died at his own hand by drinking poison, convinced of the immortality of his soul.

One of Plato's central ideas was the theory of Forms (the Greek word *eidos* is also translated as Ideas or Essences). These are ideal versions of qualities possessed by material objects to a limited and imperfect extent. They are *not* concepts but ideal objects possessing the properties to which they refer in the most perfect manner. Thus the lines and circles which we can draw are necessarily imperfect, but are approximations to the Forms of a Line and a Circle. Similar considerations apply to Beauty, Justice, and Equality. The Forms have a real, objective existence outside our minds, and our knowledge of them is acquired partly by recollection from an earlier existence, but also by disregarding the imperfect material world and withdrawing into contemplation. The role of the philosopher is to study the Forms, which alone are worthy of his attention because only they are permanent, perfect, and ultimately real.

Plato's principal use of the theory of Forms was in discussions of ethics and politics. In the *Republic* he repeatedly refers to the Forms for Justice, Beauty, and Equality:

> *Having established these principles, I shall return to our friend who denies that there is any abstract Beauty or any eternally unchanging Form of Beauty, but believes in the existence of many beautiful things, who loves visible beauty but cannot bear to be told that Beauty is really one, and Justice one, and so on,—I shall return to him and ask him, 'Is there any of these many beautiful objects of yours which may not also seem ugly? or of your just and righteous acts that may not appear unjust and unrighteous?'... Those then, who are able to see visible beauty—or justice or the like—in their many manifestations, but are incapable, even with another's help, of reaching absolute Beauty, may be said to* believe *but cannot be said to* know *what they believe.*

In mathematics abstract argument led the Greeks, and later ourselves, to an enormous flowering of knowledge, so it is understandable why Plato came to regard it as the highest type of thought. However, the development of experimental science was held back for hundreds of years by the view that the direct investigation of nature was not a suitable occupation for educated people. In the mathematical and ethical contexts Plato's theory has considerable plausibility. However, in later works Plato did not restrict the theory in this way. The following dialogue between Socrates and Glaucon in the *Republic, Theory of Art* reveals a much stronger claim about the scope of his theory.

> *'And what about the carpenter? Didn't you agree that what he produces is not the essential Form of Bed, the ultimate reality, but a particular bed?'*
> *'I did.'*
> *'If so, then what he makes is not the ultimate reality, but something which resembles that reality. And anyone who says that the products of the carpenter or any other craftsman are ultimate realities can hardly be telling the truth, can he?'*
> *'No one familiar with the sort of arguments we're using could suppose so.'*
> *'So we shan't be surprised if the bed the carpenter makes lacks the precision of reality.'*
> *'No.'*
> ...
> *'And I suppose that God knew it, and as he wanted to be the real creator of a real Bed, and not just a carpenter making a particular bed, decided to make the ultimate reality unique.'*

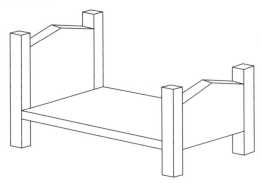

Fig. 2.1 The Carpenter's Bed

Here the process of reification is particularly clear. Plato passes from particular beds to the concept of a Bed, and then declares that there should be an ideal object corresponding to this concept. Since he is determined that this ideal object does not simply reside in our minds or our society, it must have been made by God. Plato himself was not completely happy with applying his theory of forms to material and manufactured objects, or so it appears from *Parmenides*, but it is not clear that he abandoned it.

Figure 2.1 is one of thousands of different designs for a bed, none of which can be the one made by God, but all of which are supposed to be pale reflections of the ultimate reality. I, for one, cannot imagine what the one true Bed could possibly be like, but Plato argues above that this merely proves that I do not really *know* what beds are.

The unconvincing passage about the Bed should be compared with Plato's story about the cave in the *Republic*. Here he likens non-philosophers to men imprisoned in a cave since childhood, and tied down so that they can face away from the light, so that they only see shadows of the true Reality cast onto the wall in front of them. Perhaps Plato was thinking about the problems which we discussed in the last chapter, but if so his response to them was quite different. He advocated withdrawal from the world of the senses, whereas we resort to scientific instruments to interpret it.

Another important component of Plato's philosophy is the pre-existence of the soul, discussed at length by Socrates in *Phaedo*. His argument runs as follows. We recognize that two objects are more or less equal by comparing their relationship with the Form Equality. Being parts of our material body, our senses are not capable of perceiving the Form Equality, but without an appreciation of it we could not start to make sense of the world. Therefore our understanding of it must be present at birth, and must be a memory from an earlier non-material existence. The same applies to other knowledge, which, truly speaking, is recollection from this earlier existence aided by the use of the intellect (in other places in *Phaedo* Plato seems to suggest that during abstract thought the soul can partially separate itself from the body and enter the immortal and unvarying world of Forms). From the fact that the soul pre-exists the body, we see that it is not mortal, following which a lengthy argument leads Plato to the conclusion that it is imperishable and necessarily survives death. Plato's negative view of the possibility of acquiring real knowledge in this world is made very clear in the following words of Socrates in *Phaedo*:

> *Because, if we can know nothing purely in the body's company, then one of two things must be true: either knowledge is nowhere to be gained, or else it is for the dead; since then, but no sooner, will the soul be alone by itself apart from the body. And therefore while we live, it would seem that we shall be closest to knowledge in this way—if we consort with the body as little as possible, and do not commune with it, except in so far as we must, but remain pure from it, until God himself shall release us.*

Plato's attempts to provide unassailable foundations for ethics and mathematics have been criticized from many different points of view, of which we can only select a few. The first is of a linguistic nature. Both English and Greek allow one to form abstract nouns from adjectives with great ease, but one should not suppose that by using this construction one must have identified an entity which exists independently of the language. On the contrary, an abstract noun corresponds to a concept, which might well not be associated with any type of object.[1] We saw several examples in the last chapter, such as Roman law and the ability to play a piano, whose meaning is highly culture-dependent.

A second problem concerns the uniqueness of Plato's Forms. It is certainly clear that there is only one concept of a bed, even though the boundaries of this concept are not well-defined. However, the claim that every Form is necessarily unique leads immediately to paradoxes, as pointed out by Bertrand Russell. The Form of a Triangle is a perfect triangle, so it must have three perfect edges, which are straight lines. So it seems that even God has to make three copies of the Form of a Line in order to produce the Form of a Triangle. We will discuss this in greater depth below.

Mathematical Platonism

Many mathematicians consider themselves to be mathematical Platonists in the sense described on page 27, even though they do not believe in the pre-existence of the soul, and cannot explain how one might 'see' mathematical Forms. Among the most famous of these are Kurt Gödel and Roger Penrose, both of whom made major breakthroughs in their chosen fields. I will say something about the mathematical considerations which (in my view incorrectly) led them to embrace Platonism on page 111, but let us consider their philosophical conclusions in their own right first. We start with Gödel. He believed that numbers and even infinite sets exist in themselves, and that any statement about them must be objectively true or false whether or not we know which is the case. He also believed that mathematical entities could be directly perceived:

> *But, despite their remoteness from sense experience, we do have something like a perception of the objects of set theory, as is seen from the fact that the*

> *axioms force themselves upon us as being true. I don't see any reason why*
> *we should have less confidence in this kind of perception, i.e. in mathematical*
> *intuition, than in sense perception.*

This argument is undermined by the evidence, described at some length in
Chapter 1, which establishes that our sense perceptions do *not* give us a reliable picture of the outside world. Gödel's beliefs are regarded as bizarre by
many mathematicians and philosophers, in spite of his fame. Here is a typical
comment, of Chihara:

> *Gödel's appeal to mathematical perceptions to justify his belief in sets is*
> *strikingly similar to the appeal to mystical experiences that some philosophers*
> *have made to justify their belief in God. Mathematics begins to look like a*
> *kind of theology. It is not surprising that other approaches to the problem of*
> *existence in mathematics have been tried.*[2]

Roger Penrose is even more explicit about his Platonism. The following is taken
from *The Emperor's New Mind*:

> *When mathematicians communicate, this is made possible by each one having*
> *a direct route to truth, the consciousness of each being in a position to per-*
> *ceive mathematical truths directly, through this process of 'seeing'... Since*
> *each can make contact with Plato's world directly, they can more readily*
> *communicate with each other than one might have expected ... This is very*
> *much in accordance with Plato's own idea that (say mathematical) discovery*
> *is just a form of remembering! Indeed, I have often been struck by the simil-*
> *arity between just not being able to remember someone's name, and just not*
> *being able to find the right mathematical concept. In each case, the sought-for*
> *concept is, in a sense,* already present *in the mind, though this is a less usual*
> *form of words in the case of an undiscovered mathematical idea.*

Penrose, like Gödel, seems to regard introspection as a reliable way of gaining
insights into the working of the mind. Unfortunately we have seen that twentieth
century psychological research does not support this optimism. Our impression
that we have a simple and direct awareness of the world and of our own thought
processes are both illusions. By way of contrast Einstein rejected the idea that
the nature of reality could be deduced by the application of human reason
alone:

> *At this point an enigma presents itself which in all ages has agitated enquiring*
> *minds. How can it be that mathematics, being after all a product of human*
> *thought which is independent of experience, is so admirably appropriate to*
> *the objects of reality? Is human reason, then, without experience, merely by*
> *taking thought, able to fathom the properties of real things?*
>
> *In my opinion the answer to this question is briefly, this: as far as the propos-*
> *itions of mathematics refer to reality, they are not certain; and as far as they*
> *are certain, they do not refer to reality.*[3]

He continues by contrasting the Euclidean model of reality with his own quite
different theory of relativity.

The most important contributor to the foundations of set theory since 1950 is probably Paul Cohen. In 1971 he came down decisively against Platonism (or Realism as he called it) in favour of a version of formalism. He summed up his conclusions about the possibility of *certain* knowledge in set theory as follows:

> *I am aware that there would be few operational distinctions between my view and the Realist position. Nevertheless, I feel impelled to resist the great esthetic temptation to avoid all circumlocutions and to accept set theory as an existing reality ... This is our fate, to live with doubts, to pursue a subject whose absoluteness we are not certain of, in short to realize that the only 'true' science is itself of the same mortal, perhaps empirical, nature as all other human undertakings.*[4]

Penrose's degree of commitment to Platonism is unusual, but a less explicit form of mathematical Platonism is quite common among mathematicians. On the other hand most mathematicians are painfully aware that their sudden flashes of insight are sometimes wrong, and that it is essential to check them carefully for consistency with other results in the field. Proving theorems frequently involves a high level of geometrical insight, but Penrose's idea that the sought-for concept is already present in the mind is simply wrong in many cases. If an article in a journal provides a crucial idea or technique for a theorem which you prove, it would be strange to claim that the idea of the proof was already present in your mind. It is also difficult to see what would be *meant* by saying that the concepts needed for the proof of Fermat's last theorem were already in Andrew Wiles' mind before he started trying to prove it. The fact is that theorems and/or their proofs are sometimes wrong in spite of months or years of effort on the part of their authors (e.g. Wiles' first announcement of the proof of Fermat's last theorem). This contradicts Penrose's idea that mathematicians have a direct perception of the truth. Einstein's failure to come to terms with quantum theory is another example, but in the sphere of physics rather than mathematics. The possibility of serious error also explains why mathematics journals have careful refereeing systems. If the only issue was whether the results of research papers were sufficiently important to be worth publishing, the refereeing process would be far less painful.

Penrose has replied to the above criticism by stating that it misrepresents his position.[5] He fully accepts that individual mathematicians frequently make errors, and goes on to say that he was concerned mainly with the ideal of what can indeed be perceived in principle by mathematical understanding and insight. He explains that he has been arguing that his ideal notion of human mathematical understanding is something beyond computation. Here, I am afraid, he begins to lose me. He uses the words 'ideal' and 'in principle' so frequently that I cannot relate his claims to anything I recognize in the activities of real mathematicians.

It cannot be denied that mathematical Platonism is superficially attractive as a means of explaining our intuition about natural numbers. Unfortunately the attractiveness of an idea is no guarantee of its correctness. In Chapter 3 we will see that the theory of numbers was created by society in a series of historical stages, and that we only really have a direct intuition for those at the lower end of the range. Our concept of number has been extended by the introduction of zero and negative numbers, and then real and finally complex numbers. Such changes might be explained as the result of our gaining ever clearer views of the Platonic Form of Number, but they are equally easily interpreted as the result of our *changing and developing* the concept of number in ways which we find useful or convenient. What mathematicians cannot do is accept the demise of Plato's philosophical system, and then continue to refer to it as if it were still valid.

A key issue for Platonists is the belief that any mathematical statement is true or false before anybody has determined this. Believing this, however, is not mandatory for mathematicians. Whether or not they are Platonists, everybody agrees that if a person has a genuine proof of some statement, it is not plausible that someone else will correctly prove the opposite. Issues relating to logical consistency do not only arise in mathematics. It uses such ideas more than most other fields, but they arise everywhere. For example, it is not possible that a chess player with white pieces can force a win and that the player with black pieces can do the same. Nor is it possible that I have a sibling but my sibling does not. A genuine mathematical contradiction involving the whole numbers would show that arithmetic is inconsistent. This is indeed (just about) logically possible, but it is not worth losing sleep over. If such a contradiction were to be discovered within arithmetic, it would not be a disaster, but a wonderful opportunity to look for a better theory. The experience of three thousand years shows that any such inconsistency must be very subtle, and it would not be likely to have any consequences in ordinary life.

So far I have only quoted mathematicians' views for or against Platonism. There is also a vast philosophical literature defending and criticizing Platonism in mathematics. In *Platonism and Anti-Platonism* Mark Balaguer argues as follows.[6] Human beings exist within space-time. If there exist mathematical objects then they exist outside space-time (this is what eternal means). Therefore if there exist mathematical objects, human beings could not attain knowledge of them. Balaguer then discusses at length each of the steps in this argument, concluding that only his own form of *Full-blooded Platonism* meets all the objections. Unfortunately FBP (which he is describing, not advocating) is sufficiently different from what most Platonists mean by Platonism, that it may seem that he has abandoned Platonism altogether. This impression is heightened by the fact that he finally concludes that there is no way of separating FBP from a version of anti-Platonism called fictionalism:

> *It's not just that we* currently *lack a cogent argument that settles the dispute over (the existence of) mathematical objects. It's that we could* never *have such an argument … Now I am going to motivate the metaphysical conclusion by arguing that the sentence—there exist abstract objects; that is there are objects that exist outside of space-time (or more precisely, that do not exist in space-time)*—does not have any truth condition … *But this is just to say that we don't know what non-spatiotemporal existence amounts to, or what it might consist* in, *or what it might be* like.

So in the end, the issue appears to revolve around the meaning or 'existence' or 'being'; if one adopts too simple a view of this concept, it may corrupt all of one's subsequent thought processes. We conclude with a comment of Michael Dummett, which again serves to illustrate how difficult it is to resolve such problems:

> *We do not* make *the objects but must accept them as we find them (this corresponds to the proof imposing itself upon us); but they were not already there for our statements to be true or false of before we carried out the investigations which brought them into being. (This is of course only intended as a picture, but its point is to break what seems to me the false dichotomy between the Platonist and the constructivist pictures which surreptitiously dominates our thinking about the philosophy of mathematics.)*[7]

The Rotation of Triangles

The fact that Platonic Forms are eternal by definition prevents human beings manipulating them within our experienced time. On the other hand mathematicians frequently moving their mental images around as they please. In this section we discuss an example which illustrates the difficulties which certain types of Platonist can encounter.

Let us consider two triangles, one inside the other. The bigger triangle has edge lengths 7, 8, 9, while the smaller one has edge lengths 3, 4, 5. We consider the problem

> *Can the smaller triangle be rotated continuously through 360° while staying entirely inside the bigger one?*

It may be seen from figure 2.2 that the answer is not obvious.[8]

While this appears at first sight to be a problem in Euclidean geometry, this is only true with qualifications. Euclidean geometry as described by a formal system of axioms has no notion of time. According to Plato geometry studies the properties of eternal Forms, and he dismisses the use of operational language with disdain in the *Republic*. On the other hand if one reads Euclid one finds many references to the drawing of construction lines for the purpose of providing proofs; see page 102. In his construction of spirals Archimedes refers explicitly to the passage of time and uniform rotational motion. Put briefly even the Greek geometers were not Platonists.

The problem described involves two triangles with different sizes and shapes. These may be idealized triangles, but they are certainly different ones. In this problem Platonists must concede that there is not a single Form of a triangle but at least two. Indeed there must be a different Form of a triangle for every possible size and shape. This is in conflict with Plato's insistence that there is only one Form of anything, in his discussion of beds. In this problem one is forced to concede either that the Forms of triangles may move relative to each other as time passes, or that the triangles which the mathematician is imagining are not the ideal Forms, but some other triangles which only exist in his/her

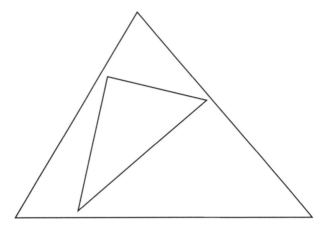

Fig. 2.2 Two Triangles

head. Plato himself struggled to find a coherent version of his theory of Forms, particularly in *Parmenides*, and never decided whether mathematical objects should be regarded as Forms or as a third class between Forms and material objects. Certainly he rejected the identification of Forms with thoughts.

Suppose that two people are asked to solve the problem at the same time. They are sure to start with the triangles in slightly different initial positions, and to rotate them in different ways, possibly in opposite directions, completing the task after a different period of time. So they cannot be imagining the same Platonic Forms, even if we concede that Forms are capable of moving as time passes. One could imagine that the entire population of the Earth was solving this problem at the same time. All of them could have the inner triangle moving in a slightly different manner. The obvious conclusion to be drawn from this scenario is that everybody is imagining a different pair of triangles. Every individual has access to a different abstract universe, populated with triangles which are capable of being moved under his/her volition. But this bears no relationship with Platonism, which supposes that Forms of triangles are independently existing motionless, ideal objects. From words he put into the mouth of Parmenides, it is clear that the later Plato was well aware of and troubled by this dilemma.

There is a way in which Platonists can try to escape from the dilemma. It depends upon declaring that the 'true' problem has nothing to do with time; the fact that we think of it that way is a consequence of our defective understanding of the Platonic reality. There are several ways of eliminating any direct reference to time. The most obvious is to introduce a third space variable, turning the original question into a problem in three-dimensional Euclidean geometry.

Specifically it asks whether there exists a solid body which satisfies certain rather strange conditions and which also can be fitted inside a cylindrical tube.

There is no doubt that any solution to the problem can be formulated in such terms. However, I consider it perverse to declare that the problem as stated, which involves moving things around on the plane, is somehow a misunderstanding of the 'true' problem. This feeling is reinforced by the fact that I cannot imagine solving the problem except in the original formulation, by trying to rotate the triangle in my own subjective time. A formalization eliminating time would not simplify or clarify the solution in any respect, and would serve only to satisfy the requirements of those who demand formality.

Most mathematicians find no difficulties with the problem as originally posed. It is theoretically possible that a non-constructive proof could exist, but it is hard to imagine what it would be like. Anybody solving the problem does so by *devising a process* for turning the smaller triangle through 360° within the larger one, and the process itself constitutes the solution. On the other hand the problem could be proved insoluble by finding a logical barrier to the existence of such a process. Indeed this example may be typical. J. R. Lucas has said, 'Mathematical knowledge is very largely knowledge how to do things, rather than knowledge that such and such is the case. Claims to know how to do something are vindicated by actually doing it.'[9]

Mathematical truth is a very slippery concept. This is not to say that it does not exist, but rather that we cannot be absolutely sure we have found it simply because we have an apparently logical proof. People make mistakes, particularly when checking a single lengthy argument repeatedly. *Our knowledge* of the truth of a mathematical statement depends upon making judgements based upon appropriate evidence. This evidence includes proofs of the type presented in text books, but may also involve numerical calculations, already solved special cases, geometrical pictures, consistency with one's intuition about the field, parallels with other fields, wholly unexpected consequences which can be verified, etc. Mathematicians try to increase their knowledge, but this knowledge is based more upon the variety of independent sources of confirmation than upon logic.

Descartes and Dualism

René Descartes (1596–1650) was one of the most important philosophical figures in Europe in the second millennium. His lasting reputation would be assured by his seminal improvements of algebra and its application to the solution of problems in arithmetic and geometry. However, he transformed many areas of philosophy in a number of books which became steadily more influential after his death. This book is not the place to celebrate his achievements, since our primary purpose is to focus on unresolved problems. Therefore we will only consider the part of his metaphysics which relates to the division between mind and body. This is widely considered to be less than compelling, in spite of its subsequent influence on the development of science.

Descartes famous aphorism 'cogito ergo sum' (I think therefore I am) and the philosophical system which he built upon it have been analysed in great detail by many scholars.[10] Its precise meaning was discussed at length by Descartes himself, and it appears that he did not consider it to be a logical deduction omitting the implied statement 'everything which thinks must exist', which would properly need to be supported by evidence. Rather he considered his thinking, his existence, and the logical connection between them all to be equally apparent to his intuition. More important is the fact that he could entertain as a *logical possibility* that the existence of the external world and even of his own body were illusions created by a deceitful spirit, whereas he could not do so with respect to his mind. Thus he came to the conclusion that mind (or soul) and body must be entirely different types of entity.

Descartes' task was then to construct an entire system of belief using rational argument starting from his 'cogito'. He recognized that reliable knowledge of the nature of the external world was extremely hard to prove, and was forced to invoke God for this purpose:

> *I had only to consider, for each of the things* of which I found some idea within me *whether it was or was not a perfection to possess the item in question, in order to be certain that none of the items which involved some imperfection were present in him, while all the others were indeed present in him ... When we reflect on the idea of God which we were born with, we see ... finally that he possesses within him everything* in which we can clearly recognize *some perfection that is infinite or unlimited by any imperfection.*

The first and weakest component of Descartes' argument is that non-existence is an imperfection, and hence existence must be among the attributes possessed by God. This is close to the so-called ontological argument of St. Anselm, which had already been rejected by St. Thomas Aquinas in the late thirteenth century in *Summa Theologica I q2*. Aquinas is quite clear that the formation of concepts has nothing to do with existence in the Platonic or any other sense:

> *Yet granted that everyone understands that by this name* God *is signified something than which nothing greater can be thought, nevertheless, it does not therefore follow that he understands that what the name signifies exists actually, but only that it exists mentally. Nor can it be argued that it actually exists, unless it be admitted that there actually exists something than which nothing greater can be thought; and this precisely is not admitted by those who hold that God does not exist.*

The second component of Descartes' argument is that God, being perfect, cannot also be deceitful. Therefore if a person has a sufficiently clear perception of some aspect of the material world, he can be confident that God would not let him be entirely misled by his senses. This leads on to his study of the nature of the world and of scientific knowledge, in which he adopts a mechanistic point of view. This was much more radical in the historical context than it might seem now. He claimed that most functions of the body did not involve the intervention of the soul, including even:

> *the retention or stamping of those ideas in the memory, the internal movement of the appetites or passions, and finally the external movements of the limbs*

which aptly follow both the actions and objects presented to the senses and also the passions and impressions found in the memory.

The behaviour of animals was entirely governed by such mechanistic processes, but Descartes did allow the human mind a limited role in acts to which we pay conscious attention:

Since reason is a universal instrument which can be used in all kinds of situations, whereas [physical] organs need some particular disposition for each particular action, it is morally impossible for a machine to have enough different organs to make it act in all the contingencies of life in the way which our reason makes us act.

'Moral certainty', he later explained, means certainty beyond reasonable doubt rather than absolute proof. The scientific philosophy of Descartes is wholly materialistic. He explained scientific phenomena by creating mechanical models to show how particles of matter interact at a scale of size which we cannot perceive directly. He countered scholastic criticisms of his approach by saying that no scientific theory could possibly be established with the same degree of certainty as theorems in geometry:

And if one wishes to call demonstrations only the proofs of geometers, one must say that Archimedes never demonstrated anything in mechanics, nor Vitello in optics, nor Ptolemy in astronomy, and so on; this, however, is not what is said. For one is satisfied, in these matters, if the authors—having assumed various things which are not manifestly contrary to experience—write consistently and without making logical mistakes, even if their assumptions are not exactly true ... But as regards those who wish to say that they do not believe what I wrote, because I deduced it from a number of assumptions which I did not prove, they do not know what they are asking for, nor what they ought to ask for.

He also emphasized the need for experimentation to distinguish between different explanations of phenomena:

I must also admit that the power of nature is so ample and so vast, and these principles so simple and so general, that I notice hardly any particular effect of which I do not know at once that it can be deduced from the principles in many different ways; and my greatest difficulty is usually to discover in which of these ways it depends on them. I know of no other means to discover this than by seeking further observations whose outcomes vary according to which of these ways provides the correct explanation.

I come now to the criticism of Descartes' philosophical system. His metaphysics has many logical flaws, which are enumerated in detail in Cottingham's anthology.[11] Even if one accepts his argument for the existence of God, only God's lack of deceit allows Descartes to be sure that his sufficiently clear beliefs guarantee correct knowledge of the material world. History shows that this is not a reliable route to knowledge. For example, possibly convinced that his own coordinate geometry was a true description of the external world, Descartes believed that he could prove by pure thought that matter must be infinitely divisible. We now accept an atomic theory of matter, and realize that one of his mistakes in this respect lay in assuming that the smallest fragments of matter must have the same character as gross matter. In fact subdividing atoms into smaller fragments *of the same type* is not just impossible but inconceivable within the conceptual framework of quantum theory. Within the pages of this book we provide many other examples of beliefs which are absolutely clear to certain groups (or at certain times) but which are equally clearly false to others. History has demonstrated time and again that Descartes' criterion of 'sufficient clarity' is so demanding that one can rarely know that it has been met. We now believe that knowledge of the material world can only be gained by testing repeatedly against experimental evidence, and that this process often leads to our having to abandon our native intuition about 'how things must be'. If he exists, God is far more subtle than Descartes imagined.

Descartes' radical separation of mind from body was extremely convenient for the development of the physical sciences, because it enabled scientists to defer indefinitely the study of minds and to declare improper any scientific reference to final causes. It came to be believed all animal motion and eventually even human behaviour were to be described in terms of the mechanical and chemical laws governing the movement of the relevant bodily parts. Even late in the twentieth century anyone who dared to diverge from this approach risked being ridiculed by 'true' scientists. Eventually the behaviourist B. F. Skinner took this idea to such extremes that a retreat was inevitable. Jane Goodall's famous study of chimpanzees in Tanzania showed that refusing to accept the relevance of goals and social relationships simply prevented one understanding their behaviour.

It is possible to argue that the development of such ideas in the seventeenth century was historically inevitable because of the advance of physical science, but Descartes was the one who articulated the ideas first. In spite of the enormous impact of Cartesianism on the development of physical science, many philosophers regard it as responsible for some our worst confusions about the nature of the world. I will pursue this issue further in Chapter 9.

Dualism in Society

Debates about whether humans or animals are conscious, have free will or have souls frequently lead nowhere, because those involved in the discussion do not realize that those to whom they are talking understand the terms differently. One of the most important analyses is due to David Hume. His first book, *A Treatise of Human Nature*, was published in 1739, and is regarded as his most important work, in spite of the fact that at the time it was almost completely ignored. He eventually rewrote a part of it as *An Enquiry concerning Human Understanding* in 1748, but this was hardly more successful at first. The *Enquiry* contained a chapter on miracles which made clear his lack of respect for religious orthodoxy, and in 1761 all of his books were placed on the Roman Catholic Index.

In the *Treatise* Hume demonstrated the possibility of discussing the nature of the will in a non-dualist framework. One of his main goals was to show that the common notion of free will put together two quite different ideas. He used the term 'will' to refer to our ability to knowingly give rise to any new motion of our body, or new perception of our mind. The word 'knowingly' is crucial to eliminate situations in which one is compelled to act as one does, or acts in the ignorance of the consequence of one's actions, or is afflicted by a serious mental incapacity (madness). He emphasized that our entire social system assumes that acts of the will are influenced by consequences to the person involved. Hume contrasted the above idea with the notion of liberty, which he considered to be either absurd or unintelligible. He argued that the possibility of making choices unconstrained by rational considerations or by passions, that is randomly, is not something to be valued. Indeed he regarded it as entirely destructive of all laws both divine and human. Thus while we can easily imagine choosing randomly to eat one piece of fruit rather than another, this says little about the human condition. On the other hand a person who made an important moral decision in a deliberately random manner in order to demonstrate their free will would correctly be regarded a suffering from an abnormal personality.

In spite of the force of Hume's arguments, they have had little influence on ordinary people, who still talk and think about free will in a dualistic manner and believe that the mind/soul is radically different from the body/matter. From the religious point of view this has the advantage of allowing the soul to continue in existence when the body has been completely consumed after death. But even the non-religious may well feel that their own subjective consciousness cannot be described in the same terms as the material world. The problems are that it seems to be impossible to say what precisely the soul is while preserving both its total distinction from the body and also its ability to interact with the body. In this section we explore a few of the religious approaches to this issue, all of which have serious deficiencies. Of course the same is true of non-religious approaches: if an approach without serious deficiencies had been discovered the problem would no longer be so contentious!

There is an important strand of Christian thinking which rejects dualism while retaining a belief in the afterlife. In *I Corinthians* Ch. 15, Paul rather enigmatically supports the idea that the soul is resurrected within a new but ideal body:

> So also is the resurrection of the dead. It is sown in corruption; it is raised in incorruption . . . It is sown a natural body; it is raised a spiritual body. If there is a natural body, there is also a spiritual body . . . Now I say, brethren, that flesh and blood cannot inherit the Kingdom of God; neither doth corruption inherit incorruption . . . But when this corruptible shall have put on incorruption, and this mortal shall have put on immortality, then shall come to pass the saying that is written Death is swallowed up in victory.

In the *Gospel of St Matthew* Ch. 13, v. 36–50 it is suggested that everyone will rise from their graves on the Day of Judgement, a notion which has been much elaborated in religious literature and art since then. Similarly the Nicene Creed of 325 AD refers to the resurrection of the body.

Unfortunately naive dualism is still alive, indeed thriving, and various cults have embraced it with disastrous results, one of the best documented being Heaven's Gate. In March 1997 a group of 37 adults committed consensual mass suicide in a mansion in San Diego, in the stated belief that they were embarking on a transfer of their minds to an extraterrestrial spaceship which was approaching the Earth behind the comet Hale–Bopp. The cult members were regarded as ordinary, non-threatening people by those who knew them; they ran a Web page design business and also used the Web heavily to publicize their views. It is clear that they were absolutely convinced that the mass of humanity were deluded about the true nature of the world, and that they themselves were going on to a higher stage of life in androgynous extraterrestrial bodies. According to the Exit Press Release of the cult itself:

> *The Kingdom of God, the Level above Human, is a physical world, where they inhabit physical bodies. However, those bodies are merely containers, suits of clothes—the true identity (of the individual) is the soul or mind/spirit residing in that 'vehicle'. The body is merely a tool for that individual's use—when it wears out, he is issued with a new one.*

The beliefs of the cult were a mixture of science fiction, mysticism, and an extreme unorthodox version of Christianity. The important point here is that their final act is incomprehensible except within a dualistic philosophy in which the mind is believed to have a separate existence from the body.

In *Beyond Science* the theoretical physicist/Anglican priest John Polkinghorne has proposed abandoning the idea of an independently exist-ing soul but within a Christian context. He instead describes the soul as the information-bearing pattern of the body, which dissolves at death with the decay of the body. He regards it as a perfectly coherent hope that the pattern will be remembered by God and recreated by him in some new environment of his choosing in his ultimate act of resurrection. There are two difficulties with this idea. The first is that it assumes that the mind remains active and clear up to the point of death. In the case of Alzheimer's victims, when is one supposed to take the pattern? If this is done before the onset of the disease then many valid experiences will be lost, but if it is taken at the point of death, almost no pattern will still exist. If the pattern is supposed to refer to the sum total of all life experiences, then it cannot be reincarnated in a body, because bodies have a location in time. The second problem is that the recreation of a person from their pattern (however that term is interpreted) cannot be regarded as the same person. If there is no physical continuity between the original and the copy, then the copy is just that. If the words 'information bearing pattern' and 'remember' have their normal meanings then one has to admit that God could make two or more such copies if he chose to do so; since it is not possible that both would be the original person, neither can be.

What can we learn from these examples? Mind-body dualism has been rejected by almost all current psychologists and philosophers on the grounds that the idea explains nothing. On the other hand many people continue to adopt a dualistic view of the world, while being deeply disturbed that groups such as Heaven's Gate or the Spanish Inquisition might actually act on that belief. The search for personal immortality is clearly a deep aspect of the human psyche, although no coherent accounts of how it could be the case have yet been produced. But the strengths of people's beliefs have not often depended on rational argument, and this one does not seem likely to be abandoned soon.

2.3 Varieties of Consciousness

We have seen that both Plato and Descartes were dualists: they believed that the soul/mind could be separated from the body/matter. Plato rejected the study of imperfect matter as worthless by comparison with his Forms, and believed that mathematical ideas had a real, independent existence which the soul could appreciate directly. Descartes had trouble explaining how minds could have reliable knowledge of the material world, and had to invoke God's help in achieving this. His philosophy of science swept all before it, but consigned the mind to an ever smaller role in the scheme of things. It now appears that many philosophers have adopted a purely materialist view in which mind is a function of or process in the brain; they regard belief in the soul as being no more than a remnant of a long outmoded system of thought.

The current debate concerns not the existence of souls but the nature of consciousness, and we will concentrate on this issue henceforth. A recent survey of some current attitudes towards the mind-body problem reveals strongly expressed disagreements on almost every issue. Current positions range from the denial of the reality of consciousness (eliminative materialism) to the statement that the solution is obvious and the suggestion that the solution is straightforward but we as humans are physiologically incapable of understanding it. An amusing comment on all this was made by the philosopher John Searle:

> *Seen in perspective, the last fifty years of the philosophy of mind, as well as cognitive science and certain branches of psychology, present a very curious spectacle. The most striking feature is how much of mainstream philosophy of mind of the past fifty years seems obviously false. I believe there is no other area of contemporary analytic philosophy where so much is said that is so implausible.*[12]

It is tempting to *define* consciousness as the ability of an entity to interact with its environment. Approaching the problem this way leads one into serious problems. While it is clear that humans and dogs are conscious, we may have legitimate doubts about ants and viruses, and few would want to allow barometers even a limited degree of consciousness. Taken literally, the definition leads us to endow everything, even atoms, with a very slight degree of consciousness, and to measure the degree of consciousness of an entity in terms of the complexity of its interactions with the environment. We are then forced to accept that computers are conscious, their 'environment' consisting of their input and output devices. This definition has the merit of simplicity and definiteness, but it trivializes many important issues relating to consciousness.

A key issue is the distinction between consciousness in the third person sense: what makes *other* people behave as they do, and consciousness in the first person sense: what is the fundamental nature of *my* subjective impressions? There seems to be an underlying dualism in the way we think about consciousness, just as there is for souls, with the difference that it is harder to deny the existence of subjective consciousness.

The difficulties of distinguishing between the two types of consciousness is demonstrated by the existence of visual illusions. We know of their existence because we experience them subjectively. On the other hand different people experience the same illusions when show the same pictures, so they also have an objective aspect. The illusions do not exist in the pictures themselves, but are produced inside our heads. We may eventually be able to explain them physically in terms of modules in our brains and unconscious processing, but this will not remove the subjective experiences.

In the remainder of this chapter we will only discuss the aspect of consciousness amenable to scientific study: third person consciousness. A variety of ideas about the nature of first person consciousness will be described in Chapter 9.

Can Computers Be Conscious?

The first goal of this section is to demonstrate that current computers are not conscious under any reasonable interpretation of the word. Then we will move into more difficult territory, with the aim of clarifying the debate rather than resolving it.

It is well recognized that computers can perform certain tasks such as the evaluation of extremely complicated numerical expressions vastly more rapidly and reliably than human beings. For certain types of algebraic mathematics computer software such as Mathematica and Maple can also out-perform us by a huge margin. However, mathematicians are not on the verge of becoming redundant! The same software packages are completely lost when faced with a problem such as proving that $n^{n+1} > (n+1)^n$ for all $n \geq 3$. The proofs of inequalities are notorious for requiring ingenuity, sometimes to an extreme degree. Nobody has yet found a means of reducing problems involving them to routine procedures which a machine could implement. I am not claiming that computers will never be able to attack such problems, but only that programs such as Mathematica and Maple are simply expert systems, provided with a set of rules by mathematicians. They are helpless in situations in which mathematicians have not been able to develop systematic procedures, even for their own use.[13]

An example of an expert system is the computer Deep Blue, the last in a series of chess-playing computers designed by IBM over a period of years. In May 1997 it played a series of six games against the world champion Gary Kasparov, acknowledged to be the greatest (human) grandmaster of all time, and beat him by $3\frac{1}{2}$ games to $2\frac{1}{2}$. Deep Blue's method of playing chess was quite different from that of a human player. It examined an enormous number of possible lines of play, using a scoring system to decide which to pursue to greater depth, and eventually choosing the optimal strategy according to rules formulated by its programmers. Its chess-playing skill came partly from its ability to examine 200 million chess positions per second, and partly from the rules programmed into it about what kind of positions to aim for and avoid. Human players, on the other hand, use an intuitive method to decide which lines of play to examine, and do not consider more than a few hundred positions in detail. I am not aware that anyone involved in the design of Deep Blue ever proposed that it was conscious or engaged in genuine thought.

The fact that Deep Blue could beat Kasparov is not really as important an issue as some people seem to think. Ten years before Deep Blue's victory chess-playing programs could already beat all but a tiny fraction of the human race. Why people should feel that their own superiority is assured if one extremely unusual individual can out-perform a computer has always been a mystery to me. The real issue is whether the processes the computer uses can be classified as conscious thinking, and in this particular case the answer is surely no.

It is interesting to consider how a human being does a computation in arithmetic. During the process we do not think about the meaning of what we are doing, but simply turn ourselves into automata, processing the data according to rules which we were taught as children. Our consciousness remains, not thinking about the *meaning* of the rules, but monitoring our successful implementation of them and checking that our attention does not wander. There are no analogous processes in a computer, since its attention cannot wander—it has literally nothing else it could be thinking about—and its level of concentration cannot vary.

Some readers may remember that the first version of the Pentium chip in the mid 1990s made occasional errors in simple arithmetic, and had to be redesigned to eliminate these. One could of course point out that humans make vastly more errors when they perform such calculations, and that we are in addition far slower. However, unlike the Pentium, we are capable of retraining ourselves if such a systematic error in our method of calculation is pointed out to us. This highly publicized accident demonstrates vividly that a computer chip is performing arithmetic mindlessly. The quality of its performance depends entirely upon its designers' care rather than on its own abilities to think through problems.

On the other hand, one can compare a pocket calculator with a mobile telephone. In spite of the fact that each can do things quite beyond the capacities of the other, they are about the same size, both have keyboards, both have LED displays and both have computer chips inside. Their different capacities result from different internal organizations of their components, and the ability of the mobile telephone to transmit and receive messages. So it may be with human beings. The internal architecture of our brains is totally different from that of computers, and we have the advantage that enormous amounts of information flood into our brains through our sense organs constantly. In some respects we out-perform computers and in others they out-perform us. That does not prevent both computers and ourselves being finite computing machines. Whether this is, in fact, the case is another matter. The best way of determining the answer is to try to understand *how* we think and copy it on a suitable machine.

A similar issue arises in comparing us with eagles. As far as flying is concerned they win hands (or wings) down, but when using screwdrivers we outperform them almost as dramatically. That does not imply that there is some deep difference in our cellular structure, just that it is organized differently. A moment's thought about the differences between animals shows how import- ant the organization of cells/genes is to the properties of the final creature. We are said to have over 98% of our genes in common with chimpanzees,[14] but even this small difference has led to remarkable differences in our intellectual capacities. We should therefore keep an open mind about the possibility that rad- ically different hardware or software could change our view about the potential capacities of computers to think in the sense we normally use this word.

Gödel and Penrose

Kurt Gödel's importance in the foundations of mathematics, discussed in Chapter 5, is so great that it compels us to listen to his comments on the differences between human thought and that of computers. Since his views on some key issues were diametrically opposed to those of Alan Turing, almost equally important in this field, we cannot however simply defer to his author- ity. When one looks at what he has written, much of it seems curiously out of tune with the current views of both philosophers and scientists, who often have good reasons for not understanding what he is trying to say. At the very least his views are unfashionable, but the grounds for rejecting them should be stated explicitly. Hao Wang has reported on a number of discussions with Gödel in the early 1970s, paraphrasing his views as follows:

> *Even if the brain cannot store an infinite amount of information, the spirit may be able to. The brain is a computing machine . . . connected with a spirit. If the brain is taken as physical and as a digital computer, from quantum mechanics there are then only a finite number of states. Only by connecting it to a spirit might it work in some other way . . . The mind, in its use, is not static but constantly developing . . . Although at each stage of the mind's development the number of its possible states is finite, there is no reason why this number should not converge to infinity in the course of development.*[15]

Wang tried to disentangle the utterances Gödel was inclined to make into the defensible and those which are essentially mystical. The first two sentences of the quotation embrace dualistic thought in a way which is rare in other philosophical writings of the twentieth century. The meaning of the last sentence is extremely unclear. While nobody would argue with the mind being an open-ended learning system, it obviously cannot literally acquire an infinite amount of knowledge. The fact that a person might in principle acquire an infinite amount of knowledge if he/she were to keep on learning for an infinite length of time has no consequences in the real world. If Gödel simply means that during an actual finite life span a person might keep on learning new facts, ideas, and techniques, then his use of the word 'infinity' can only serve to confuse.

At earlier periods in his life Gödel expressed quite different views; for example in his 1951 Gibbs lecture he stated:

> On the basis of what has been proved so far, it remains possible that there may exist (and even be empirically discoverable) a theorem-proving machine which in fact is equivalent to mathematical intuition, but cannot be proved to be so, nor even be proved to yield only correct theorems of finitary number theory.

Here we have Gödel accepting that human mathematical abilities may be capable of being matched by a machine. The only problem is that neither the machine nor the mathematician would then be using provably correct algorithms. In contrast to this, let me quote from an article of Penrose published in 1995:

> Are we, as mathematicians, really acting in accordance with an unconscious unknowable algorithm? One inference from such a proposal would be that the reasons we offer for believing our mathematical results are not the true reasons for such belief. Mathematics would depend upon some unknown calculational activity of which we were never aware. Although this is possible, it seems to me unlikely as the real explanation for mathematical conviction. We have to ask ourselves how this unconscious unknowable algorithm, of value only for doing sophisticated mathematics, could have arisen by a process of natural selection.[16]

My colleague Larry Landau has recently given a careful response to this highly controversial argument.[17] Brains work by trying to create patterns (i.e. mental models) which match what they learn from the outside world or from introspection. This process is almost entirely unconscious and cannot be categorized as logically sound or unsound, because it is just the operation of a physical mechanism. The procedures which we use consciously when doing mathematics are entirely different from the mechanisms which control the functioning of our brains. If our conscious thought processes are occasionally or even systematically unsound, this has no implications concerning the mechanisms used by our brains to produce these thoughts. There is even an aphorism which fits this situation: do not blame the messenger for the message.

One of the popular approaches to artificial intelligence uses the theory of neural networks. Scientists in this field try to model our brains by constructing machines which learn by experience. In a very narrow sense their operation is algorithmic, in that the machines are electronic computers running programs. On the other hand the machines *teach themselves* how to recognize individual patterns. Often they get the wrong answer, but as time passes the frequency of mistakes decreases. The performance of such machines is far below what we achieve, but the machines are far simpler than our brains, so this is not surprising. Such machines function in a way which Penrose above considers to be unlikely as a model for our own thinking, but many others consider the analogy very convincing.

This idea about how the brain works provides an answer to Penrose's question about how our mathematical ability could have evolved. Pattern recognition is *the primary ability* of our brains, and the development of this ability over millions of years is what has made us what we are. The co-option of this ability for mathematical purposes did not need any further evolution, but depended upon the social environment becoming suitable for such activities. No special algorithms for doing mathematics exist, and no guarantees of correctness of the insights obtained are available.

Discussion

Over the last decade the introduction of a variety of scanning machines has led to a revolution in the understanding of consciousness. These machines allow researchers to watch the activity in different parts of people's brains as they are set various tasks. This is one of the most the most exciting current area of scientific research, but that does not mean that it is near to solving the main problems. The human brain is incredibly complicated, and ideas are in a constant state of flux, with the eventual conclusions by no means clear, even in outline. I am not the person to review progress in this highly technical field, and will only try to describe certain issues which the final theory will have to explain. Even this is a daunting task.

In scientific publications consciousness almost always refers to the third person or 'objective' sense of the word, that is to some unknown aspect of the neural mechanisms people and animals use when they direct their attention to a matter requiring decision making. How might we decide whether an animal is conscious in this sense? The first possibility is that we observe their behaviour, and agree to call them conscious if this is sufficiently close to our own. Let us look at it in more detail.

There have been vigorous arguments among biologists about whether complicated goal-directed behaviour among higher mammals is reliable evidence for their consciousness. Indeed the admission of consciousness into animal research is quite a recent phenomenon. Injury-avoidance behaviour is often based on reflexes, and it is not completely obvious that the inner sensation of pain must be attached to it. Even in our own case pain is often felt only *after* the limb has been moved away. Again, many birds build sophisticated nests entirely instinctively, and may or may not be conscious of what they are doing. At the other end of the animal kingdom octopuses and squid have entirely different brain anatomies from ourselves and our common ancestor probably had no brains at all. Nevertheless they are capable of learning and memorizing facts for months. If they are to be included in the realm of conscious beings, this indicates that consciousness does not depend upon a particular type of brain anatomy.

Recent investigations of the behaviour of bees when choosing a new home are particularly interesting as a test of what we mean by consciousness.[18] A swarm of bees often rests in a tree after leaving its original hive, while a small number of scout bees look for possible new homes. The studies show that the decision process does not depend upon any of the scouts visiting more than one site. On this return each scout signals some characteristic of a site to the swarm by a special dance. No bee has enough brain capacity to assess the relative merits of the sites, and the decision is taken by a group process which does not require any member of the swarm to be aware of the whole range of possible new sites. Nor does it require the scouts or any other individual bees to take a final decision on behalf of the swarm. To call the swarm conscious *as a swarm* would be rash indeed when we have no detailed understanding of consciousness even for humans, let alone for other organisms.

On the other hand it seems impossible to describe the behaviour of such a swarm without referring to goals. The above paragraph uses the words 'choosing', 'look for', and 'decision'. Perhaps *we* need to use such teleological language for what are actually purely instinctive responses, because this is the only way we can relate to the physical behaviour of swarms. This problem recurs throughout the biological sciences.

Returning to human beings, the distinction between conscious and unconscious behaviour is a real one. When we learn to drive a car, we are initially highly conscious of every action needed. By the time we become experienced drivers the mechanical aspects of driving have moved to the periphery of our attention and it is possible to conduct a conversation at the same time as driving. Very occasionally our attention may slip entirely and we may experience the sudden shock of realizing that we have no memory of the last traffic lights which we passed. This indicates that behaviour in humans becomes conscious not because it is complex, but when it involves unfamiliar or deliberate choices.

Consider next the process of breathing. Mostly it is not under our conscious control, and even if we run to catch a bus we do not make a conscious choice to increase the rate and depth of our breathing. However, it is possible for us to take over conscious control of our lungs for short periods, and there can be no doubt that when we do this something different is happening in our brain than when we breathe unconsciously. Although it is impossible for us to tell by observing animals in the field whether they are able to control their breathing consciously, it is clear from our own case that this is a real question, and not one about the use of words.

A further proof that consciousness cannot be identified with behaviour comes by considering those unfortunate people who suffer total paralysis, while retaining their mental faculties. On fortunately rare occasions this happens during surgical operations, when patients are mistakenly given the usual muscle relaxants but without sufficient anaesthetics.[19] The unfortunate patients are in no doubt that their induced paralysis is totally different from anaesthesia.

We have already discussed the phenomenon of blindsight, which illustrates well the difference between consciousness and the ability to process information. There is, however, absolutely straightforward evidence that much of our thinking is unconscious and inaccessible. This is the process of remembering facts which are not near the front of one's mind. Many readers will remember occasions on which they wanted to remember the name of someone they had not seen for several years, or to recall a word in some foreign language which they once knew. It is possible to spend several seconds, or, as you get older, even minutes, trying to remember the required word, and then for it to pop into your mind without warning. There is something going on in one's brain, and it is quite sophisticated since it involves the meanings of words. Nevertheless we have no idea how our minds are obtaining the required information, nor where they got it from when it arrives.

One cannot simply dismiss unconscious thought as referring to low level processes. Creative thinking involves unconscious processes which are capable of solving problems which our conscious minds cannot. When thinking about an intractable problem it is common for mathematicians (and others!) to spend months trying all the routine procedures, and then put the problem aside. Frequently a completely new idea comes to them suddenly, in a flash of insight similar to that which supposedly came to Archimedes in his bath. As a typical example of this process consider Hamilton's account of his discovery/construction of quaternions[20] in 1843, following fifteen years of unsuccessful attempts:

> *On the 6th day of October, which happened to be a Monday, and Council day of the Royal Irish Academy, I was walking to attend and preside, and your mother was walking with me along the Royal Canal; and although she talked with me now and then, yet an undercurrent of thought was going on in my mind, which gave at last a result, whereof it is not too much to say that I felt at once the importance. An electric current seemed to close; and a spark flashed forth ...*

In *Science and Method*, 1908 Henri Poincaré wrote in almost identical terms about flashes of insight he obtained during his study of the theory of Fuchsian functions. He also mentioned that these flashes were not invariably correct, as did Jacques Hadamard in *The Psychology of Invention in the Mathematical Field*, 1945. In some way prolonged and unsuccessful attempts to solve a problem stimulate one's unconscious mind to search for new and fruitful lines of attack. When something seems (to the unconscious mind) to have a high probability of leading to the solution, it forces the idea to the mathematician's conscious attention. The criteria used by the unconscious mind are certainly not trustworthy, and the ideas so obtained have to be checked in detail. On the other hand the phenomenon often achieves results which conscious, rational thought cannot.

For the above reasons, I take it as established that the distinction between conscious and unconscious thought is a matter of fact: we do not just attach the label consciousness to all sufficiently high level processes in our brains out of convention. Consciousness does not imply the ability to perform actions, and does not arise in many types of sophisticated brain activity. There is a specific brain mechanism whose activation causes conscious awareness. We do not yet know what this mechanism is or how it interacts with other parts of the brain, but it is likely that this will be elucidated over the next twenty years.

We now come to a further difficulty. Human beings have a higher type of consciousness than any other animals. Only humans and the great apes can recognize that the images they see in mirrors are of themselves and that unusual features such as marks seen on their foreheads might be removed. Only in their fourth year do children start to recognize the possibility of false beliefs in themselves and others, to remember individual events for long periods and to be able to make complex plans for the future. There is an enormous research literature on the ways in which our thought processes differ from those of all other animals, and the stages at which our special abilities develop during childhood.

Episodic memory, the ability to transfer individual thoughts to and subsequently from memory, is generally agreed to be of vital importance for higher level consciousness. One possibility is that higher level consciousness arises within a yet to be located physical module in the brain which deals with this ability. A person is conscious when this module is acting normally, while dreaming might correspond to a different mode of action of the module. The *routine* exercise of skills such as riding bicycles, walking, driving, etc. does not pass through it, but when something unexpected happens the decisions made are routed by the module into memory, from which they can then influence subsequent behaviour.

Some support for the above idea may be found in the recent research literature. Eichenbaum states:

> The hippocampus is crucial for memory and in humans for 'declarative memory', our ability to record personal experiences and weave these episodic memories into our knowledge of the world around us.[21]

If only matters were so simple! In a recent article John Taylor pointed out the problems with all current proposals for the seat of consciousness, and eventually came down in favour of the inferior parietal lobes.[22] We are evidently still far from being able to identify the hypothetical module, and it remains possible that consciousness cannot be localized in this manner. It may correspond to some characteristic wave of activity sweeping through the entire brain. Whatever the situation, suppose brain scientists succeed in identifying a neural mechanism (a module or mode of activity) which corresponds perfectly to our own subjective experience of consciousness: when it operates we are conscious and when it is disabled by brain damage or anaesthetics we are not. From the point of view of the physiologist this would be a satisfactory solution of the problem of consciousness.

Suppose next that we can design a machine which has suitably rich inputs from and outputs to the external world and which contains a hardware or software implementation of the neural mechanisms in our own brains. Would it then be conscious in the proper subjective sense? Or would it merely be simulating consciousness? Perhaps this is incapable of being decided in an absolute sense, but we would be forced to treat such machines as conscious. The point is that the reasons for believing them to be conscious would be as good as the reasons for believing other people to be conscious. The only 'reason' for believing them not to be conscious would be the fact that they were built in a factory rather than grown inside someone's womb.

The development of machine consciousness may fail purely because of the difficulty of the project. The neural network approach asserts that a brain is essentially a set of neurons, and that we can duplicate that function of the brain without worrying about other aspects. This leads to two problems. The first is that there are many key aspects of brain function which are not neural, depending upon complex chemical messengers such as endorphins and neuropeptides. The second is that neurons can grow new connections in response to external stimuli and internal injuries. The way in which they 'know' where to grow is under genetic control in a general sense, because everything is, but it also depends heavily on other factors which we hardly understand at all.

Whether it may one day be possible to produce a neural analogue of the brain depends upon the nature of its organization. A human brain has about 10^{11} neurons, each possessing around ten thousand synapses, far more than any electronic machine at the present time. There is no evidence that it is possible to copy the activity of the brain in some device which is radically smaller. As soon as one looks at animals whose brains are ten times smaller than ours, one sees that all of our advanced skills have disappeared. Even chimpanzees, which have brains with about one-quarter of our number of neurons, cannot begin to match our mental achievements. It is of course possible that the organization of our brains is so inefficient that one could achieve as much with a very much smaller number of efficiently arranged components. If this is the case it is at worth asking why we have not evolved such a more efficient brain, since its energy requirements impose a very heavy burden upon us: although it weighs less than two kilos, about 20% of our energy is devoted to keeping it running. This figure is so large that there must have been a very large evolutionary incentive for our brains to become more efficient.

The main hope of success of the research programme depends on the brain being massively redundant, so that a much smaller and simpler system may still capture its essential features. We do not know whether this is so, but the effort of finding out will surely teach us an enormous amount. The interaction between neuro-anatomists and those involved in the design of neural networks promises to bring understanding and possible treatment of mental disorder whether or not it leads to a proper model of consciousness. One should not underestimate the magnitude of the task. Our thinking appears to be controlled by intuitive judgements about whether the procedures we are adopting are appropriate to the goals which we set ourselves, rather than by logical computation. We do not understand how to design a computer program which could copy this behaviour, even with the recent advances in neural network theory. To declare that all will become clear with a few more years' research is to make a declaration of faith rather than a sober assessment of the current position.

Let us again suppose that there is a specific brain mechanism which is involved in conscious behaviour in humans. This mechanism must be linked to a large number of modules relating to motor functions and acquired skills. There is now experimental evidence that the precise forms and even locations of these modules vary from person to person, and these will influence the expression of the consciousness mechanism. As we learn more about people's brains and how they develop, the notion of consciousness (in the third person sense) will become much more precise and complicated. Whether this will also solve the problem of subjective consciousness is a matter of debate, as we will see in Chapter 9.

Notes and References

[1] In the scholastic tradition the view I am advocating is called conceptualism, and is contrasted with Plato's realism—which I prefer to call Platonism.

[2] Chihara 1990, p. 21

[3] Einstein 1982a

[4] Cohen 1971

[5] See Penrose 1996, 6.2.1 This article also contains links to a number of articles by his critics, several also writing in 'Psyche'.

[6] Balaguer 1998, p. 22

[7] Dummett 1964, p. 509

[8] In fact it is not quite possible.

[9] Lucas 2000, p. 366, 367

[10] Cottingham 1992

[11] Cottingham 1992

[12] Searle 1994

[13] This may change with the development of 'genetic algorithms', but it is too early to say how far this idea will go.

[14] Recent studies suggest that this figure should not be relied on.

[15] Wang 1995, p. 184

[16] Penrose 1995, p. 25

[17] Landau 1996

[18] Seeley and Buhrmann 1999, Visscher and Camazine 1999

[19] In 1999 anaesthetists did not have a guaranteed method of measuring the depth of anaesthesia.

[20] See page 71 for further details.

[21] Eichenbaum 1999

[22] Taylor 2001

3
Arithmetic

Introduction

Much of modern science depends on the heavy use of mathematics, justly known as the Queen of the Sciences. Let me list just a few of its many achievements. Euclid's geometry was for two millennia the paradigm of rigorous and precise thought in all other sciences. The introduction of the Indo-Arabic system of counting and the logarithm tables of Napier and Briggs early in the seventeenth century were vital for the development of navigation, science and engineering. Newton was forced to present his law of gravitation in purely mathematical terms: he tried and failed to find a physical mechanism which would explain the inverse square law. Darwin's epochal theory of evolution was entirely non-mathematical, but the recent development of genetics and molecular biology have been heavily dependent on the use of mathematical techniques. The two technological revolutions of the twentieth century, quantum theory and computers, have both been highly mathematical from their inception. Indeed the two pioneers in the invention of computers in the 1940s, Alan Turing and John von Neumann, were both mathematicians of truly exceptional ability.

The success of mathematics in so many spheres is a great puzzle. Why is the world so amenable to being described in mathematical terms? Albert Einstein described this as the great puzzle of the universe, joking that 'God is a mathematician', and many distinguished scientists have echoed this sentiment. Both mathematicians and philosophers have believed at various times that our insights into Euclidean geometry, Newtonian mechanics, and set theory/logic are exempt from the general limitations on human knowledge. Unfortunately each has eventually been proven not to have any such status; we will see how this happened later on. In this chapter I explain why our concept of number is also not nearly as simple as most people consider. A long historical process has resulted in the creation of a powerful structure which we now use with confidence. But this must not blind us to the fact that the properties of numbers which we regard as self-evident were not always so.

For a mathematician to cast doubt on the independent existence of numbers might seem bizarre. Read on, and I will try to persuade you that this is actually irrelevant to the pursuit of mathematics. What we actually depend upon is a set of rules for producing theorems, together with informal procedures for generating intuitions about those rules. It would be psychologically convenient if the rules concerned were properties of some external set of entities. Many mathematicians behave as if this were the case. But it is not necessary. The meanings of road signs are entirely conventional, but they nevertheless explain a lot about the flow of traffic. Nobody suggests as a result that green→go and red→stop are fundamental laws of nature. In both cases one can simply accept that if these are the rules, then those are the consequences.

You might also ask, if someone believes that numbers do not exist independently of ourselves, why do they bother to pursue the study of their properties? The answer is the same as might be given by musicians and artists. The process of creation, and of appreciation, gives enormous pleasure to those involved in it, and if others also find it worthwhile, so much the better.

Whole Numbers

For the purposes of this discussion we will divide numbers (natural numbers, positive integers) into four types according to the following rules:

> *one to ten thousand—small*
> *ten thousand to one trillion—medium*
> *one trillion to 10^{100}—large*
> *much bigger than that—huge*

Here a trillion is a million million and 10^{100} is 1 followed by 100 zeros. I do not insist on the exact boundaries between these ranges, and would accept, for example, that the small numbers might include everything up to a million. However, there are real differences between the four ranges, which I need to describe in order to set the scene for the arguments below. The above separation of numbers into different types is similar to the division of colours according to their various names: the fact that the categories overlap and that people may disagree about the borderlines does not make the distinction a worthless one.

I will not plunge straight into asking what numbers really are, since this might rapidly become either a technical discussion in logic or a philosophical debate[1]. Perhaps examining the history of counting systems will prove a more enlightening way into the subject.

Small Numbers

These are numbers which one uses in ordinary life to count objects. For example ten thousand represents the number of points in a square of 100×100 points. It is also the number of steps which one takes in a walk of two to three hours. Our present notation for counting in this range functions so smoothly that we can easily forget the history behind its development.

In Roman times the numbers 5, 50, and 500 were represented by different symbols, namely V, L, and D. Instead of having separate symbols for each number from 1 to 9, they used combinations of a smaller number of symbols. Since this system is now almost entirely confined to monuments recording births and deaths of famous people, I summarize its structure. The symbols used in what we call the Roman system are

$$I = 1 \quad V = 5 \quad X = 10 \quad L = 50 \quad C = 100 \quad D = 500 \quad M = 1000.$$

The integers from 1 to 10 are represented by the successive expressions

$$I \quad II \quad III \quad IV \quad V \quad VI \quad VII \quad VIII \quad IX \quad X$$

those from 10 to 100 in multiples of 10 by

$$X \quad XX \quad XXX \quad XL \quad L \quad LX \quad LXX \quad LXXX \quad XC \quad C$$

and those from 100 to 1000 in multiples of 100 by

$$C \quad CC \quad CCC \quad CD \quad D \quad DC \quad DCC \quad DCCC \quad CM \quad M.$$

Thus the date 1485 is represented by MCDLXXXV, the fact that C is before D indicating that the C should be subtracted rather than added. The final year of the last millennium is MCMXCIX; a simpler but less systematic notation is MIM.

The system described above is only one of a number of variations on a common theme. In medieval times a wide range of notations was used. One convention put a line over numbers to indicate that they were thousands, so that $\overline{IV}CLII$ would represent 4152. Another was to separate groups with different orders of magnitude by dots, so that II.DCCC.XIIII would stand for 2814. The Romans themselves used the symbol ⅭⅠↃ to represent 1000, and the right hand half of this, D, came to represent a half of a thousand, namely 500.

The tomb of Galileo Galilei in Basilica di S. Croce in Florence records his year of death as CⅠↃ.ⅠↃ.C.XXXXI. This is somewhat puzzling, since he died in 1642, not 1641, by our calendar. The explanation is that in Florence at that time the year started on 25 March, and Galileo died in January. This is only one of several problems one has to confront when converting dates to our present calendar; another is that historians often do not indicate whether they have carried out the conversion or not, leading to further scope for confusion!

The task of multiplying two Roman-style numbers is not an easy one. Consider the following apparently very different formulae, which use the seventeenth century multiplication sign.

$$IV \times IX = XXXVI$$
$$XL \times IX = CCCLX$$
$$IV \times XC = CCCLX$$
$$XL \times XC = MMMDC$$
$$IV \times CM = MMMDC$$
$$CD \times IX = MMMDC$$

From our point of view these all reduce to $4 \times 9 = 36$, with zeros attached in various places. In the Greek and Roman worlds people had to learn a new set of tables for each order of magnitude; the procedures were sufficiently complicated that whole books such as Heron's *Metrica* were devoted to what we now regard as routine arithmetic. An alternative was to turn to the use of abacuses, which were certainly well known in classical Greece, and may well be of Babylonian origin. Archimedes devoted his book *The Sand-Reckoner* to the description of a very complicated system for representing extremely large numbers.

The far better Hindu-Arabic system of counting was committed to writing by Al-Khwarizmi between 680–750 AD. It is, however, certainly *much* older than that. It came into use gradually in Europe between 1000 AD and 1500 AD, and was eventually to sweep everything else away. Its superiority relied ultimately upon the Hindu invention of a symbol for zero. The importance of this is that one can distinguish between 56 and 506 or 5006 by the presence of the zeros, and so does not need to have different symbols for the digits depending on whether they represent ones, tens, hundreds, etc. We now take all of this for granted, but the systematic use of zero was a long drawn out process, whose impact may be as great as that of any other single idea in mathematics.

Medium Numbers

The number one trillion is so big that it is not possible for one to reach it by counting. To prove this I describe a way in which one cannot get seriously rich. Suppose you could persuade someone to give you every dollar bill which you could mark a cross on. You settle down to marking one bill per second and decide to work a ten hour day. After one day you have earned $36,000 and after a working month of 25 days you have accumulated $900,000. At the end of your first year you have approximately $10 million. At this rate you would take 100,000 years to reach a trillion dollars. Even if you could speed the process up a hundred times you will still get nowhere near a trillion dollars in your lifetime. The only way to accumulate this sort of money is to emulate Bill Gates or become the dictator of a very wealthy country.

There are, however, ways in which one can make the number easier to imagine. A one kilogram bag of sugar contains about one million grains, each about one millimetre across. To acquire one trillion grains of sugar one therefore needs a million bags, which occupy about a thousand cubic metres. This would fill all of the space in two or three of the semi-detached houses of the London street in which I live.

In spite of their enormity, such numbers are important in our modern world, since several national economies have GNPs of this order. This was not always the case—when I was a boy a British billion was what we now call a trillion, and the conflict with USA usage hardly mattered because numbers of this size almost never arose. Perhaps one reason for their recent importance is that our computers can count up to these numbers even if we cannot.

Large Numbers

When we turn to scientific problems, we routinely find it necessary to go far beyond the limitations of medium numbers. Examples are the number of hydrogen atoms in a kilogram and the number of neutrinos emitted by a supernova. Hindu mathematicians had words representing some very large numbers, but until the sixteenth century there was no *systematic notation* for writing them down.

The invention of the power notation opened up the possibility of describing very large numbers in a compact notation. We write 10^m to stand for the number which we would otherwise write as 1 followed by m zeros, and 3.4×10^{54} to stand for 34 followed by 53 zeros. One must not be misled by the simplicity of this notation. 10^{54} is not just a bit bigger than 10^{51}—it has three extra zeros and is a thousand times bigger!

A few examples of the power of this notation is in order. One of the notable astronomical events in recent years was the observation of a supernova exploding in 1987. In truth it exploded 166,000 years ago, but for all of the time since then the light informing us of that fact has been making its way towards us. Its distance in kilometres is

$$(3 \times 10^5) \times 60 \times 60 \times 24 \times 365.25 \times 166000 = 1.6 \times 10^{18}.$$

The size of a measured number clearly depends upon the physical units chosen, light-years or kilometres in the above example. When we talk about large numbers below, we refer to whole numbers, all of the digits of which are significant, not to measured quantities, for which only the first few digits are likely to be accurate. The distinction between the three categories of number is easy to grasp visually. One can write down a typical randomly chosen number in each of the first three categories as follows:

1528 is small,
4852060365 is medium,
56457853125600322565752345019385012884720337503463 is large.

We have a completely unambiguous way of representing large numbers, and can distinguish between any two of them. We also have ways of adding and multiplying two large numbers, by a scaled up version of the rules we are taught in our primary schools. In other words we can manipulate large numbers satisfactorily, although they do not retain the same practical relationship with counting as small numbers do.

What Do Large Numbers Represent?

It is now time to discuss the relationship between counting and the natural world. I claim that large numbers are only used to measure quantities. More precisely there are *no situations in the real world* in which large numbers refer to counted objects.

Let us start with the number of people in the world at the instant when the second millennium ended. This was about 8 billion, so it is a medium number in my system of classification. Nevertheless even this number is difficult to define precisely, let alone evaluate. People are born and die over a period which may last from a few seconds up to several hours. There is no way of specifying either of these processes sufficiently precisely for us to be able to define a moment at which they might be considered to happen. It follows that the number of people in the world at any moment has no precise value.

Another example is the number of trees in a wood. Here the problem is what constitutes a tree. In addition to well-established trees several decades old, there will be saplings at all stages of growth, down to seeds which have only just started germinating. The point at which one decides whether or not to include something as a tree may affect the total by a factor of two or more. Of course one may make an arbitrary decision, such as requiring the height of a tree to be at least one metre, but this will still leave marginal cases. Even if it happens not to, there is no particular merit in using this way of defining tree-hood.

With genuinely large numbers the situation is far worse. Let us think about the number of atoms in a cat. One can estimate this by weighing the cat and estimating the proportions of atoms of the different chemical elements, but this is not counting. If we insist on an exact answer and the cat had a meal a few hours ago, do we regard the meal as a part of it, or when does it become a part? The cat breathes in and out, leading to a constant flow of oxygen and carbon dioxide atoms in and out of its lungs. At what exact stage are these regarded as becoming or no longer being a part of the cat? Clearly these questions have no answers, and the number of atoms in the cat has no exact meaning.

This problem cannot be avoided in any real situation involving large numbers. In an attempt to find a physical example which involves a precisely defined large number, let us consider the atom-counting problem for air sealed inside a metal box. In this case the number of atoms is not well-defined because of the process of diffusion of gas through the walls of the box. How far into the walls of the box should an oxygen atom diffuse before it is regarded as a part of the walls rather than a part of the gas? Of course, one can *imagine* a perfectly impermeable box with a definite number of atoms inside it, but this then turns into a discussion of idealized objects rather than actual ones.

The conclusion from considering examples of this type is that large numbers *never* refer to counting procedures; they arise when one makes measurements and then infers approximate values for the numbers. The situation with huge numbers, defined below, is much worse. Scientists have no use for numbers of this next order of magnitude, which are only of abstract interest.

Addition

The notion of addition is more complicated than we normally think. There are in fact two distinct concepts, which overlap to a substantial extent.

Suppose we are asked to convince a sceptic that $4 + 2 = 6$. We would probably say that 4 stands for four tokens | | | | as a matter of definition, that 2 stands for two tokens | | and that addition stands for putting these groups of tokens together thus creating | | | | | |, which is six tokens. A similar argument could be used to justify the sum $13 + 180 = 193$, but now we would have to give rules for the interpretation of the composite symbol 13 as | | | | | | | | | | | | | with similar but very lengthy interpretations of 180 and 193 as strings or blocks of tokens.

This is not, however, the way in which anyone solves such a problem. We learn tables for the sums of the numbers from 1 to 9 and also quite complicated *rules* for adding together composite numbers (those with more than one digit). Most people make the step between the two procedures for addition so success-fully that they forget that there is a real distinction. However, when one sees the difficulties a young child has learning the rules of arithmetic, it is obvious how major a step it is.

To develop the point, suppose we wish to add the number

$$3141592653589793238462643383279502884197169399937510$$

to itself. The rule-based approach to addition can be applied without any trouble. (At least in principle. It might need several attempts to be sure you have the right answer.) The problem is that it is very hard to argue that this is still an abbreviation of a calculation with tokens, which cannot possibly be carried out. It is all very well to say that it can be carried out in principle, but what does this actually mean if it cannot be carried out in the real world? Edward Nelson and others have advocated the idea that the addition of very large numbers means no more than the application of certain rules. This is an 'extreme formalist position' in the sense that it depends upon viewing the arithmetic of very large numbers as a game played with strings of digits rather than an investigation of the properties of independently existing entities. The rules are not arbitrary: they developed out of the idea of putting tokens together in groups of ten and then groups of a hundred, and so on. But eventually the system of rules took over until for large enough numbers that is all that is left. The tokens have disappeared from the scene, since we cannot imagine huge numbers of them with any precision.

To summarize, when adding small numbers we can use tokens or rules, and observe that the two procedures always give the same answer. This fact is not surprising because the rules were selected on the basis of having this property. However for large numbers one can only use the rules. In the shift from small to large numbers a subtle shift of meaning has occurred, so that for large numbers the *only* way of testing a claimed addition is to repeat the use of the rules. The rules are exactly what computers use to manipulate numbers. We like to feel that we are superior to them because we understand what the manipulations *really mean*, but our sense of superiority consists in being able to check that the two methods of addition are consistent for small numbers.

Multiplication

In *Shadows of the Mind* Roger Penrose claimed that we can see that

$$79797000222 \times 50000123555 = 50000123555 \times 79797000222$$

without performing the two multiplications, as follows. Each side of the equation represents the number of dots in a rectangle whose sides have the appropriate lengths. Since the one rectangle is obtained by rotating the other through 90° they must contain the same number of points. Penrose states that

> we merely need to 'blur' in our minds the actual numbers of rows and columns
> that are being used, and the equality becomes obvious.

Notice that the matter only becomes 'obvious' by blurring the numbers. This is necessary since the numbers are so large that they cannot be represented by rows of dots in any real sense. One can argue that the process involved is not one of perception but one of analogy with examples such as $6 \times 8 = 8 \times 6$, for which Penrose's argument is indeed justified. The analogy depends upon the belief that 6 and 79797000222 are the same type of entity, when historically the latter was obtained by a long process of abstraction from the former.

The fact that the product of two numbers does not depend upon the order in which they are multiplied is called the commutativity identity. Symbolically it is the statement that

$$x \times y = y \times x$$

for all numbers x and y. Its truth is clear provided the numbers are small enough for us to be able to draw the rectangles of dots. The question then becomes whether we can *extend* the notion of multiplication to much larger numbers in such a way that the identity remains valid. It may be shown that this is achieved by using the familiar rules for multiplication for large numbers.

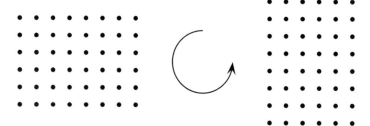

Fig. 3.1 Multiplication Using Rectangles

It may also be proved using Peano's postulates for huge numbers (discussed below). Having found an extension of the notion of multiplication which retains its most desirable properties, eventually we decide that the extension *defines what is meant* by multiplication, and forget the origins of the subject.

There is another reason for doubting Penrose's explanation of why we believe the commutativity law. In order to explain this I need to refer to entities called complex numbers. In the sixteenth century Cardan and Viète showed that certain calculations in arithmetic could be carried out more easily by introducing nonsensical expressions such as $\sqrt{-5}$, ignoring the fact that negative numbers do not have square roots. It was repeatedly observed that if one used square roots of negative numbers in the middle of a calculation but the final answer did not involve them, then the answer was always correct! It was later realized that all of the paradoxes of this subject could be reduced to justifying the use of the *imaginary* number

$$ i = \sqrt{-1}. $$

In 1770 Euler wrote in *Algebra*:

> Since all possible numbers that can be imagined are either greater than or less than or equal to zero, it is evident that the roots of negative numbers cannot be counted among all possible numbers. So we are obliged to say that there are impossible numbers. Hence we have had to come to terms with such numbers, that are impossible by their very nature and which, by habit, we call imaginary because they only exist in the imagination.

Clearly he did not subscribe to the belief that complex numbers existed in some Platonic realm.

The first stage in the demystification of complex numbers was taken by Argand and Gauss around 1800 when they chose to represent the complex number $x + iy$ (already assumed to exist in some sense) by the point in the plane with horizontal and vertical coordinates (x, y). The paradoxical square root of minus one was then represented by the point with coordinates $(0,1)$.

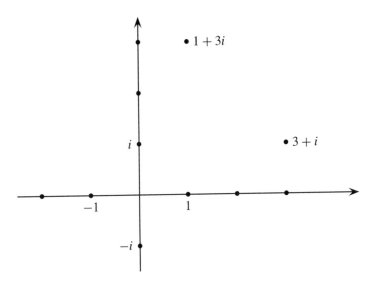

Fig. 3.2 The Complex Number Plane

In 1833 Hamilton approached complex numbers the other way around. He *defined* a complex number to be a point on the plane, and then *defined* the addition and multiplication of such points by certain algebraic formulae. Following this he was able to prove that addition and multiplication had all of the properties we normally expect of them with the additional feature that $i = (0, 1)$ satisfies $i^2 = -1$. So within this context -1 does indeed have a square root!

Hamilton's work was revolutionary because it forced mathematicians to come to terms with the fact that truth and meaning depend on the context. Within the context of ordinary (real) numbers -1 does not have a square root, but if the meaning of the word number is extended appropriately it may do so. The same trick is used in ordinary speech. Throughout human history it was agreed that humans would like to fly but could not. Now we talk about flying from country to country as if this is perfectly normal. Of course we have not changed, but we have redefined the word 'fly' so that it can include sitting inside an elaborately constructed metal box. As a result of extending the meaning of the word something impossible becomes possible.

It is extremely hard for us to put ourselves in the frame of mind of Euler and others, who could not believe in complex numbers but could not abandon them either, because of their extraordinary usefulness. Psychologically the acceptance of complex numbers came when mathematicians saw that they could *construct* complex numbers using ideas about which they were already confident. This was a revolution in mathematics, which involved abandoning the long-standing belief that mathematics was the science of magnitude and quantity.[2] It opened up the possibility of changing or extending the meaning of other terms used in mathematics, and of creating new fields of study simply by declaring what the primary objects were and how they were to be manipulated. In this respect mathematics is now a game played according to formal rules, just like chess or bridge.

The system of complex numbers is enormously useful, and mathematicians now feel as comfortable with them as they do with ordinary numbers. The important point for us is that the multiplication of complex numbers is commutative. The only reason for the truth of the commutativity identity $z \times w = w \times z$ for complex numbers is that one can evaluate both sides of the equation using the definition of multiplication and see that it is indeed true. Hamilton's subsequent invention of quaternions in 1843 was an even more revolutionary idea. These were also an extension of the concept of number, but in this case allowing the possibility that $z \times w \neq w \times z$. The technicalities need not concern us, the crucial point being that nobody previously had thought that the commutativity of multiplication was among the things a mathematician might consider giving up. Hamilton's conceptual breakthrough led to Cayley's systematic development of matrix theory in 1858 and Clifford's introduction of his Clifford algebras in 1878; in both of these the commutativity of multiplication was abandoned.

The importance of these ideas can hardly be exaggerated. If one had to identify the two most important topics in a mathematics degree programme, they would have to be calculus and matrix theory. Noncommutative multiplication underlies the whole of quantum theory and is at the core of some of the most exciting current research in both mathematics and physics.

We conclude: the fact that multiplication of ordinary and complex numbers are commutative has to be proved, and this is possible as soon as one has written down precise definitions of the two types of number and of multiplication. The fact that the multiplication of quaternions or matrices is not commutative is equally a matter of proof. Reference to the rotation of rectangles may be used to persuade children that the commutativity property is true for small numbers, but it does not provide a proper (non-computational) proof for all numbers.

Inaccessible and Huge Numbers

There are known to be infinitely many numbers. The argument for this is simple—there cannot be a biggest number, since if one reached it by counting, then by continuing to count one finds larger numbers. For numbers radically bigger than 10^{100}, however, we have no systematic method of doing computations. By radically bigger I do not mean 10^{1000}, even though it is certainly vastly bigger than 10^{100}. Nor do I mean

$$2^{13,466,917} - 1$$

which has just established a new record as the largest known prime. This number has just over four million digits, and would fill a very large book if printed out in the usual decimal notation. Incidentally the proof that this number is indeed a prime took 30,000 years of computer processing time.

By huge I mean a number such as $10^{10^{100}}$, which has 10^{100} digits. The task of writing down the digits of a typical 'randomly chosen' number with 10^{100} digits in the usual arabic notation is beyond the capacity of any conceivable computer—it would take all of the atoms in the Universe even if one could store a trillion digits on each atom. Nor can one carry out arithmetic with such numbers: the problem

$$8^{8^{888}} + 9^{9^{999}} = ?$$

just sits there mocking our impotence.

A lot is known about prime numbers both theoretically and computationally. To illustrate the latter aspect, there are exactly 9592 primes with five or fewer digits, the smallest being 2 and the largest being 99991. In order to illustrate the failure of systematic computation for huge numbers, let us define P to be the number of primes which have fewer than a trillion digits. Quite a lot is known about the distribution of prime numbers, and the prime number theorem allows one to evaluate P quite accurately. On the other hand everything we know about prime numbers suggests that the question

Is P an even number?

will never be answered. A Platonist would regard this as having an entirely straightforward meaning, but this does not help him or her one iota to determine the answer. In fact a Platonist is no more likely to solve this problem than a mathematician who regards proving things about huge numbers as a formal game.

I frequently hear mathematicians saying that questions such as the above pose no problem 'in principle'. This phrase makes me quite angry. It might mean 'I know it is not actually possible but would like to close my mind to this fact and pretend that I could do it if I really wanted to'. Another possible meaning is 'I do not regard the difficulty of carrying out a task as an interesting issue'. Either interpretation leaves the speaker cut off from the mainstream of human activities. The second amounts to a rejection of all matters associated with numerical computation, a subject which contains many challenging and fascinating problems.

Even if one has a formula for a particular huge number one may not be able to answer completely elementary questions about it. Consider the famous Fibonacci sequence

$$1, 1, 2, 3, 5, 8, 13, 21, 34, 55, 89, 144, \ldots$$

The rule for generating this sequence is that each term is the sum of the previous two. Letting f_n denote the nth term one may compute

$$f_{100} = 354224848179261915075$$

$$f_{1000} = 434665576\dots6849228875$$

$$f_{10000} = 336447648\dots947336875.$$

The number f_{1000} has 209 digits while f_{10000} has 2090 digits! Now let us now consider f_n for $n = 10^{10^{100}}$. From a naive point of view there appears to be no difficulty in knowing what we mean by this number: one just keeps on adding for an extremely long time, using an amount of paper which steadily increases as the numbers get bigger. Unfortunately in practice there appears to be no way of determining even the first digit of this number. The exact definition and the known analytic formula for f_n are equally powerless to help us, because the numbers involved have so many digits.

Some numbers, such as $10^{10^{100}}$, are perfectly simple to write down in spite of being huge. The following argument shows that most are completely inaccessible. One may classify the complexity of a number in terms of its shortest description. Thus

31415926535897932384626433832795028841971693993751 0

is lengthy when written down as above, but has a much simpler description as the integer part of $\pi \times 10^{50}$. Defining the complexity of a number in terms of its shortest possible description is fraught with problems if expressed so briefly, because of phrases such as

> The smallest number whose definition requires at least a million symbols.

Such a number cannot exist, since the above phrase defines it in 73 symbols including spaces. The accepted way out of this self-reference paradox is to replace it by the phrase

> The smallest number whose definition using the programming language X requires at least a million symbols.

Here X could be Java, C^{++}, some extension of these which permits strings of digits of arbitrary length to represent numbers, or any other high level programming language. The number defined depends on the programming language used, but for our purposes the important issue is that one does not get trapped by the illogicalities of natural language. Using this definition of complexity we see that $10^{10^{100}}$ is very simple, since it requires only 13 symbols when written in the form 10^{10^{100}}, which C^{++} is able to understand.

Some truly enormous numbers can be written down within the above constraints. For example we can put

$$a = 9$$
$$b = 9^a$$
$$c = 9^b$$
$$d = 9^c$$

without beginning to approach the self-imposed constraints on size.

Standard high level programming languages allow one to go far beyond this by means of the definition

```
n:=1;
for r from 1 to 100 do
n:=n^n + 1;
end;
N:=n;
```

For those who do not feel at ease with computer programs, it describes a procedure for generating numbers starting from 2. The next number is $2^2 + 1 = 5$. The third is $5^5 + 1 = 3126$. The fourth number in the list, namely $3126^{3126} + 1$, is still just small enough to be calculated by current PCs: it has 10,926 digits and can be printed out on about ten pages of A4 paper. The fifth number is too large for any computer constructible in this universe to evaluate (in the usual decimal notation). Only an insignificant fraction of the digits in the answer could be stored even if one allocated a trillion digits to every atom in the universe. The hundredth number in the list, which we call N, is mind-bogglingly big, and little else can probably be said about it.

Writing down the shortest description of a number is quite different from representing it by a string of digits. In spite of this, the following technical argument shows that using shortest descriptions does not materially alter how many numbers can be written down explicitly. We consider a programming language which uses a hundred types of symbol, including letters, numbers, punctuation marks, spaces, and line breaks. Suppose we consider numbers whose definition can be given by a program involving no more than a thousand such symbols. The total number of 'programs' which we can write down by just putting down symbols in an arbitrary order is vast, but it can be evaluated by using a coding procedure. We first list the hundred symbols in some order, putting the numbers from 01 to 99 and finally 00 underneath them. The list might start with:

q	w	e	r	t	y	u	i	o
01	02	03	04	05	06	07	08	09

p	a	s	d	f	g	h	j	k
10	11	12	13	14	15	16	17	18

Then we replace each symbol in the program with the number underneath it. So 'queasy' would be replaced by 010703111206. The result is to replace every program by a number with at most 2000 digits, so the total number of such programs is 10^{2000}. Actually almost all of them are gibberish, so the number of grammatical, or meaningful, programs is very much smaller. Programs do indeed allow one to write down a few numbers which are ridiculously large, but they do not provide a *systematic* way of writing down all such numbers.

The conclusion is that whether we define numbers by strings of digits, or by descriptions in a chosen programming language, there is little difference in how many numbers we can effectively express. Neither approach overcomes the basic information theoretic problem that there are limits on how many different numbers we can hope to write down explicitly, and therefore to what we can actually compute.

Peano's Postulates

In everyday life we constantly rely upon the idea that if two events have regularly been associated, they will continue to be so. Thus we 'know' that if we bring our hands together sharply, we will hear a clapping noise. We believe this not because we know anything about the physics of sound production, but simply on the basis of experience. In Chapter 9 I will discuss Hume's criticisms of this type of induction, but here I wish to consider what is called mathematical induction.

This is *not* the very dangerous idea that you are justified in believing a statement such as

For every positive number n the expression $n^2 + n + 41$ is prime.

simply by testing it for more and more values of n until its repeated validity persuades you of its general truth. The first few values of the above 'Euler polynomial' are

$$43, 47, 53, 61, 71, 83, 97, 113, 131, 151, 173, 197, 223, \ldots$$

which are certainly all prime numbers. After testing several more terms one might easily come to the conclusion that the claim is true. Actually it is false, the smallest value of n for which the expression is not prime being $n = 40$.

In order to *prove* statements about *all* numbers, of whatever size, mathematicians use abstract arguments based on Peano's postulates for the integers.

Peano wrote down his postulates (or axioms) in 1889. They are

0 is a number.
For every number n there is a next number, which we call its successor.
No two numbers have the same successor.
0 is not the successor of any number.
If a statement is true for 0 and, whenever it is true for n it is always also true for the successor of n, then it is true for all numbers.

The critical axiom above is the last, called the principle of mathematical induction. It is usually written in the more technical and compressed form

If $\mathcal{P}(0)$, and $\mathcal{P}(n)$ implies $\mathcal{P}(n + 1)$, then $\mathcal{P}(n)$ for all n.

Here $\mathcal{P}(n)$ stands for a proposition (statement) such as

$$(n + 1)^2 = n^2 + 2n + 1$$

or

Either n is even or $n + 1$ is even.

The principle of induction is not a recipe which solves all problems about numbers effortlessly, but it is the first thing to try.

A present-day Platonist might say that the truth of Peano's postulates can be seen by direct intuition. In *Science and Hypothesis*, 1902 Henri Poincaré adopted the Kantian view that 'mathematical induction is imposed on us, because it is only an affirmation of a property of the mind itself'. These arguments were not accepted by Bertrand Russell and other logicians early in the twentieth century who tried to construct the theory of numbers from the more certain and fundamental ideas of set theory. If one looks at the historical record, Russell's caution about the 'obviousness' of induction is certainly justified. The Greeks did not use it, although it is possible to detect hints of such ideas in a few isolated texts.[3] Its first explicit use in mathematical proofs is often ascribed to Maurolico in the sixteenth century. Even today, many mathematics students who have been taught the principle are reluctant to use it, preferring to rely upon direct algebraic proofs of identities if they can.

Alternatively we may regard Peano's postulates as a system of axioms like any other. That is they are a list of rules from which interesting theorems may be proved. These theorems agree with what we know for small and medium numbers because we can see that those satisfy the stated postulates—except that the obviousness of the last one becomes less clear as the numbers increase, and lose their obvious connection with counting. If the historical record is a guide, Peano's axioms are less obvious than those of Euclidean geometry. They may be seen as a part of a formal system of arithmetic. For an entertaining account of the complexities involved in setting up such a formal system see *Gödel, Escher, Bach: An Eternal Golden Braid* by Douglas Hofstadter.

Paul Bernays considered that 'elementary intuition' faded out as numbers become larger:

> *Arithmetic, which forms the large frame in which the geometrical and physical disciplines are incorporated, does not simply consist in the elementary intuitive treatment of the numbers, but rather it has itself the character of a theory in that it takes as a basis the idea of the totality of numbers as a system of things as well as of the idea of totality of the sets of numbers. This systematic arithmetic fulfils its task in the best way possible, and there is no reason to object to its procedure, as long as we are clear about the fact that we do not*

take the point of view of elementary intuitiveness but that of thought formation, that is, that point of view Hilbert calls the axiomatic point of view . . . However, the problem of the infinite returns. For by taking a thought formation as the point of departure for arithmetic we have introduced something problematic. An intellectual approach, however plausible and natural from the systematic point of view, still does not contain in itself the guarantee of its consistent realizability. By grasping the idea of the infinite totality of numbers and the sets of numbers, it is still not out of the question that this idea could lead to a contradiction in its consequences. Thus it remains to investigate the question of freedom of contradiction, of the 'consistency' of the system of arithmetic.[4]

Even if we adopt the first position (naive realism), we have to admit that for huge numbers, Peano's postulates provide the only route to our knowledge of them—the only way of convincing a sceptic that a claim about huge numbers is true makes use of Peano's postulates or something which follow from them. The following example illustrates the issues involved.

The statement that 8 is a factor of $9^n - 1$ means that if one divides $9^n - 1$ by 8 then there is no remainder, or equivalently that

$$9^n - 1 = 8 \times s$$

for some number s. A geometrical proof of this statement for $n = 2$ can be extracted from Figure 3.3. A geometric proof is also possible for $n = 3$ by decomposing a $9 \times 9 \times 9$ cube in a similar fashion.

For $n = 4$ one may check that

$$9^4 - 1 = 8 \times (1 + 9 + 9^2 + 9^3)$$

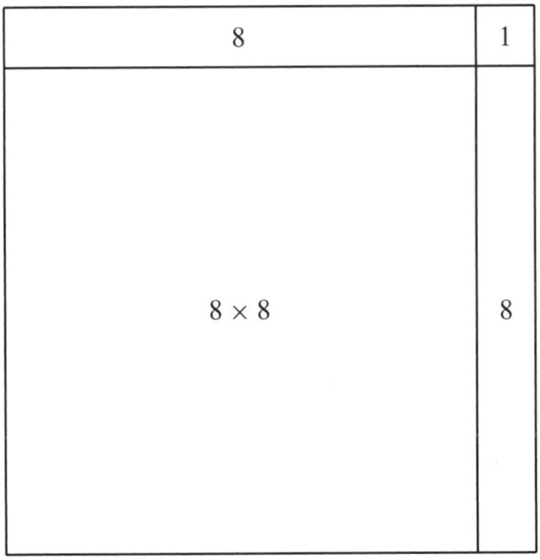

Fig. 3.3 Decomposition of a 9×9 Square

by explicitly evaluating both sides. This correctly suggests the general formula

$$9^n - 1 = 8 \times (1 + 9 + 9^2 + \cdots + 9^{n-1})$$

for all numbers n. However, this expression cannot be checked directly for $n = 10^{100}$ because the number of additions involved would take impossibly long. Also the formula involves the mysterious ... which invites one to imagine doing something, and should not be a part of rigorous mathematics. A more formal expression would be

$$9^n - 1 = 8 \times \sum_{r=0}^{n-1} 9^r$$

in which the summation symbol $\sum$ is given a formal meaning by means of the Principle of Induction. So eventually one has to believe that the use of this Principle is permissible in order to prove that 8 is a factor of $9^n - 1$ for *all* n.

One now comes to the philosophical divide. A Platonist believes that the Principle of Induction is a true statement about independently existing objects. The alternative view is that mathematicians are investigating the properties of systems which we ourselves construct, what Bernays called thought formations. Motivated by our intuition of small numbers, we decide to include the Principle of Induction among the rules which we use to prove theorems. Theorems correctly proved within such a system are true, because truth is always understood as relative to some agreement about the context.

In this view mathematics is not like exploring a country which existed long before the explorer was born. It is more like building a city, with its unlimited potential for muddle, error, and growth. We lay the foundations of each building as well as we can, but accept the possibility of collapse. If a building does fall down, we rebuild it, learning from our errors. We also examine other buildings to see if they have the same design faults. Gradually the city becomes more impressive and better adapted to our needs, but it always remains our creation.

There still remains something to be said about Peano's Principle. If one is not willing to declare that its truth is self-evident, how can one justify its use for large numbers, the ones which scientists have a real use for? In linguistic terms, if we define large numbers by the strings of digits used to manipulate them (their syntax), then we have removed the only reason for believing Peano's postulate, which is based on the meaning of number (their semantics). This seems a fatal blow to formalists who would argue that large numbers are *no more than* long strings of digits. Fortunately one can prove Peano's Principle for large number strings in a few lines using only conventional logic. This resolves the objection.[5]

Infinity

If one can entertain doubts about the meaning of very large finite numbers, then it seems that we have no hope of understanding the infinite. My intention here is to persuade you that there are many different meanings to 'infinity', written as ∞, all of which are of value in the appropriate context. Each of them captures some aspect of our intuitive ideas about the infinite, which, as finite beings, we cannot perceive directly. In Chapter 5 we will discuss whether infinite objects actually 'exist', and what this might mean.

The obvious way of dealing with infinity is to write down rules for manipulating it, such as $\infty \times \infty = \infty$ and $1 + \infty = \infty$ and $1/\infty = 0$. One quickly finds that great caution is needed in manipulating such algebraic expressions. Otherwise one may obtain nonsensical results such as

$$0 = \infty - \infty = (1 + \infty) - \infty = 1 + (\infty - \infty) = 1 + 0 = 1$$

or

$$1 = \frac{\infty}{\infty} = \frac{2 \times \infty}{\infty} = 2.$$

Nevertheless infinity is used in this manner by all analysts, who learn to avoid the pitfalls involved.

There is a different way of introducing infinity, which is quite close to the modes of thought in the subject called non-standard analysis. Namely one introduces a symbol ∞ and agrees to manipulate expressions involving it according to the usual rules of algebra. In this context $\infty \times \infty$ is not merely different from ∞, but vastly (indeed infinitely) bigger. Similarly $1/\infty$ is not equal to 0 but it is an unmeasurably small number, called an infinitesimal. Finally ∞ is bigger than every positive integer, but $\infty + 1$ is bigger than ∞. It can be shown that if one follows certain simple rules there is no inconsistency in this system, which does capture some of the properties which we think infinity should have. Note, however, that the two notions of infinity above are different and incompatible with each other. Which we decide to use depends upon what we want to do.

The symbol ∞ also appears as a shorthand for statements which avoid its use. Thus writing that something converges to 0 as n tends to infinity is just another way of writing that it gets smaller and smaller without ceasing. The corresponding formal expression

$$\lim_{n \to \infty} a_n = 0$$

means neither more nor less than

$$\forall \varepsilon > 0. \exists N_\varepsilon . n \geq N_\varepsilon \to |a_n| \leq \varepsilon.$$

One need not understand either of these formulae to see that the infinity in the first has miraculously disappeared in the second, being replaced by the logical symbols $\exists$, $\forall$, $\to$. The credit for providing this rigorous 'infinity free' definition of limit goes to Cauchy in *Cours d'Analyse*, published in 1821. The symbol ∞ is considered to have no meaning in isolation from the context in which it appears. Analysts agree that this type of use of the symbol does not involve any commitment to the existence of infinity itself.

The above notions of infinity provide more precise versions of previously rather vague intuitions. Since there are several intuitions one ends up with several different infinities. The above is typical of how mathematicians think: we start from vague pictures or ideas, about infinity in this case, which we try to encapsulate by rules, and then we discover that those rules persuade us to modify our mental images. We engage in a dialogue between our mental images and our ability to justify them via equations. As we understand what we are investigating more clearly, the pictures become sharper and the equations more elaborate. Only at the end of the process does anything like a formal set of axioms followed by logical proofs appear. Eventually we come to behave as if the ideas which we have reached after much struggle already existed before we formulated them. Perhaps later generations do not realize that other ideas were pursued and abandoned, not because they were wrong but because they were less fruitful.

Discussion

The division of numbers into small, medium, large, and huge was a device used to focus attention on the fact that successive stages of generalization involve losses as well as gains. At one extreme numbers really do refer to counting, but at the other the relationship with counting only exists in our imagination.[6] The most abstract, and recent, concept of number depends upon formal rules of logic and Peano's property. By distinguishing between these four different types of number I seem to be violating the principle of Ockham's razor:

non sunt multiplicanda entia praeter necessitatem

i.e. entities are not to be multiplied beyond necessity. The following are some positive reasons for distinguishing between the types of number. The fact that computers can manipulate large numbers with great efficiency, but are pretty hopeless beyond that, suggests fairly strongly that huge numbers are genuinely different from large ones. In algorithmic mathematics the size of the numbers involved in a procedure is one of the primary issues, and the appearance of huge numbers indicates that the procedure is not of practical use. Abstract existence proofs often provide little information about the properties of the entity proved to exist; often they do little more than motivate one to seek to find more direct computational methods of approach which provide more information about the solutions.

A version of the following paradox was known as the 'Sorites' in the Hellenistic period, but in the form below it is due to Wang.[7] It states

The number 1 is small.
If n is small then n + 1 is small.
Therefore, by induction, all numbers are small.

It is as clear to philosophers as to others that the conclusion of this argument is incorrect, so the only issue can be to explain where the error lies. You may ask why anyone should bother about such a trivial issue. The answer is that there may be other arguments which are incorrect for the same reason, even though in these other cases it may not be at all obvious that an error has been made. The same applies to the exhaustive enquiries held after a crash of a commercial airliner. They cannot save the life of anyone who died, but, if the reason for the crash is discovered, it may be possible to prevent it happening again.

Michael Dummett has discussed this paradox at length and raised doubts about whether one can apply the normal laws of logic to vague properties such as smallness.[8] There are in fact (at least) two ways of resolving such problems, both of which would be perfectly acceptable to any mathematician, if not to philosophers. The simplest is to simply declare numbers less than a million (say) to be small and others to be big. Dummett mentions this possibility but declares it to be a priori absurd; he ignores the fact that this is precisely the way in which the law distinguishes between children and adults, another vague issue. An alternative is to attach an index $s(n)$ of smallness to every number, by a formula such as

$$s(n) = \frac{10^6}{n + 10^6}.$$

Using this formula, numbers which we think of as small get a smallness index close to 1, while very large numbers get an index close to 0. The particular formula above assigns the smallness index $0 \cdot 5$ to the number one million, so if one uses this formula one would regard a million as being intermediate between small and large. We could then say that the common notion of smallness merely attaches the adjective to all numbers for which the speaker considers the index to be close enough to 1, but all precise discussions should use the index. Either of these proposals immediately dissolve the paradox. There is even a mathematical discipline which studies concepts which do not have precise borderlines, called fuzzy set theory.

The status of the Peano property is different for numbers of each size. For 'counting' numbers its truth is simply a matter of observation. For numbers defined as strings of digits I have shown how to prove it in a recent publication. For huge or formal numbers it is an abstract axiom. Each of the three ways of looking at numbers has its own interest, and one learns valuable lessons by finding out which problems can be solved within each of the systems. We should distinguish between the features of the external world which lead one to some idea (counting in the present context), and the mathematical system we have invented to extend that initial idea (formal arithmetic). Historically it is clear that our present idea of number is far removed from that of our ancestors, and that we may never have a good way of handling most huge numbers.

The belief that individual numbers exist as objects independent of ourselves is far from being accepted by philosophers. Paul Benacerraf has examined in detail a number of different ideas about what numbers might be if they do exist, coming to the conclusion:

> *Therefore numbers are not objects at all, because in giving the properties (that is, necessary and sufficient) of numbers you merely characterize an abstract structure—and the distinction lies in the fact that the 'elements' of the structure*

have no properties other than those relating them to other 'elements' of the same structure. … Arithmetic is therefore the science that elaborates the abstract structure that all progressions have in common merely in virtue of being progressions. It is not a science concerned with particular objects—the numbers.[9]

Let me give a brief flavour of his argument. The number 'three' may be represented by the symbols *III* or 3. One may construct the number using the supposedly more fundamental ideas of set theory in at least two different ways. All of these methods of expressing numbers yield the formula $4 + 1 = 5$, sometimes as a theorem and sometimes as a definition of 5 or of +. Similarly with other rules of arithmetic. There seems to be no way of persuading a sceptic that any of these expressions for the number is more fundamental than any other. Benacerraf concludes that 'three' cannot *be* any of the expressions, and that one can use any progression of symbols or words to develop an idea of number.

Let us nevertheless concede for the moment that small or 'counting' numbers exist in some sense, on the grounds that we can point to many different collections of (say) ten objects, and *see* that they have something in common. The idea that the Number System as a whole is a social construct seems to lead one into fundamental difficulties. I will examine these one at a time.

If one is prepared to admit that 3 exists independently of human society then by adding 1 to it one must believe that 4 exists independently. Continuing in this way seems to force the eventual conclusion that $10^{10^{100}}$ exists independently. But as a matter of fact it is not physically possible to continue repeating the argument in the manner stated until one reaches the number $10^{10^{100}}$. We must not be misled by the *convention* under which mathematicians pretend that this is possible 'in principle'.

If one does not believe that huge numbers exist independently then how can they have objective properties? The answer to this question is similar to that for chess. Constructed entities do indeed have properties, and while some of these may just be conventions, others may not be under our control. We decide on rules which we will obey in chess, and then we play according to the rules. Our agreement about the truth of theorems is of the same type as the agreement of people in the chess world about the correctness of a solution to a chess problem. One difference between mathematics and chess-players is that mathematicians are constantly altering the rules to see if we can find more interesting games. However, mathematicians may only change the rules within certain conventionally prescribed limits or they are deemed no longer to be doing mathematics. For example they cannot make the rules depend upon whether or not it is a religious holiday. Nor can they make the rules depend upon the geographical location of the practitioner, as can lawyers.

Pure mathematicians reject issues relating to religion, race, nationality, gender, and even views about the structure of the world as valid bases for arguments, so it is not entirely surprising that they have been able to achieve a considerable consensus on the very rarified world remaining. Once one progresses sufficiently far in the creation of any social structure, whether it be mathematics or law, it takes on a life of its own, dictating what can and cannot be done. Every now and again controversies arise even in mathematics, but they may be examined for years or even decades before a consensus emerges. Even then the issues involved may be raised again if it appears worthwhile to do so; as time passes the task becomes ever greater because of the accumulated work based on the dominant tradition.

There is a final question. If one does not believe that many of the entities in mathematics have an independent existence, how does one account for the extraordinary success of mathematics in explaining the world? There cannot be a simple answer to this question, to which we will return in the concluding section of the book. One part of the answer is that we understand the universe to the extent that we can predict its behaviour. Our 'extraordinary success' is only extraordinary by standards which we ourselves have set. We need to keep reminding ourselves that there exist chaotic phenomena which we will never be able to predict whether or not we use mathematical methods. Our own existence, both as a species and individually, depends upon historical contingencies whose details could not possibly be explained mathematically. While new scientific theories will certainly be developed, we do not expect these to be able to bypass the above problems. Quantum theory indicates that at a small enough scale prediction is fundamentally impossible, except in a probabilistic sense. We will see some of the evidence which justifies these claims in later chapters.

Notes and References

[1] A similar division was described by Paul Bernays [Bernays 1998], in an article discussing the philosophical status of numbers in 1930–31. It is also possible to provide an empiricist defence for the existence of such a division. See Davies 2002 and Gillies 2000a.

[2] Dunmore 1992, p. 218

[3] Acerbi 2000

[4] Bernays 1998, p. 253

[5] Davies 2002

[6] In technical terms I am suggesting a realist ontology for small numbers and an anti-realist ontology for large enough numbers.

[7] See Acerbi 2000 for references to the literature on the 'Sorites'.

[8] Dummett 1978, p. 248–268

[9] Benacerraf 1983, p. 291. Benacerraf himself and several other philosophers criticized the argument of this paper subsequently. See Morton and Stich, 1996.

4
How Hard can Problems Get?

Introduction

The stock portrait of a pure mathematician is of a thin, introverted person, who is socially inept and likes to sit alone contemplating unfathomable mysteries. There is more than a grain of truth in this image. On the other hand I know mathematicians who are continuous balls of energy. A few have acquired the status of prophets in their own lifetimes, and are regularly surrounded by rings of disciples. Some have long term goals towards which they direct their energies for years on end. Yet others spend their lives hacking through a jungle, hoping to find something of interest if only they persist for long enough.

The one thing which unites all these different people is incurable optimism. Not that this is obvious! Gödel proved that there are mathematical problems which are insoluble by normal methods of argument, but all mathematicians are sure that their own particular concern does not fall within this category. Indeed they have immense faith that if they persist long enough they will surely make some progress in resolving the issue to which they are devoting their energies.

Roger Penrose based his popular books on the argument that while Gödel's theorem constrains computing machines, human beings can transcend its limitations. Put briefly they can 'see' the truth without the need for chains of logical argument. To explain this he postulates that microtubules in neurons allow the influence of quantum effects on conscious thought. I do not have the expertise to judge whether microtubules and consciousness have some deep connection, and am happy to leave time to judge that issue. I am, however, less happy with Penrose's belief that human beings have potentially unlimited powers of insight. Indeed it strikes me as astonishing, since all of our other bodily organs have obvious limits on their capacities. However, what different people do or do not find incredible is of less significance than what we discover when we look at the evidence.

In this chapter I describe a few of the outstanding mathematical discoveries which have taken place during the last half century. They were not selected randomly: each of them says something about how far human mathematical powers extend. This, rather than their mathematical content, is also what I concentrate on when discussing them. Together they suggest that we are already quite close to our biological limits as far as the difficulty of proven theorems is concerned. This should not be taken as indicating that mathematics is coming to an end. New fields are constantly opening up, and these always start with ideas which are much more easily grasped than those of longer established fields. It is likely that interesting new mathematics will continue to appear for as long as anyone can imagine, because we will constantly discover new *types* of problem. This, however, is quite a different issue.

When mathematicians talk about hard problems, we may mean one of several things. The first relates to problems which are hard in the mundane sense that great ability and effort are needed to find the solutions. The second sense is more technical and will be explained in the section on algorithms. There are finally statements which are undecidable within a particular formal system in the sense of Gödel; we will not discuss Gödel's work extensively since much (possibly too much) has already been written about the subject.

The remainder of this chapter may be omitted without serious loss. The topics which I have chosen are completely independent, and you are free to read any or all as you wish. (There is no examination ahead!)

Before considering very hard problems let us look at one of intermediate difficulty. Pure mathematics and in particular arithmetic are often said to be *a priori* in the sense that the truths of theorems do not depend upon any empirical facts about the world. It is sometimes said that even God could not stop the identity $3^2 + 4^2 = 5^2$ from being true! In 1966 Lander and Parkin discovered the identity

$$27^5 + 84^5 + 110^5 + 133^5 = 144^5$$

by a computer search,[1] thus disproving an old conjecture of Euler that the equation

$$a^5 + b^5 + c^5 + d^5 = e^5$$

has no solutions such that a, b, c, d, e are all positive whole numbers. The solubility of this equation is an example of an *a priori* fact. On the other hand it has a definite empirical tinge, in the sense that the solution was only discovered by a computer, and verifying that it is indeed a solution would involve about six pages of hand calculations. I know of no proof of solubility which provides the type of understanding a mathematician always seeks, and there is no obvious reason why a simpler proof should exist.

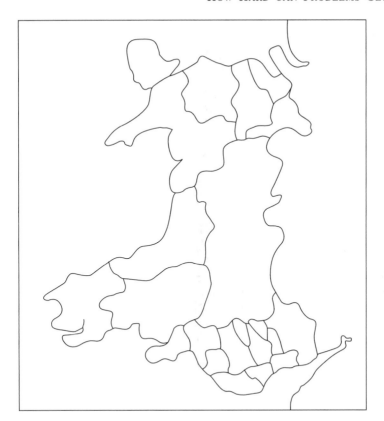

Fig. 4.1 The Welsh Local Authorities

The Four Colour Problem

The four colour problem concerns the number of colours needed to cover a large plane area divided up into regions (a map) in such a way that no two neighbouring regions have the same colour. The conjecture is (or rather was) that four colours suffice for any conceivable map.

The problem was formulated by Guthrie in 1852, and over the next hundred years a number of incorrect proofs of the conjecture were found. In 1976 Appel and Haken used a combination of clever mathematical ideas with lengthy computer calculations to provide a genuine proof. There were some blemishes in their first published solution, but these were later corrected.

Their method could not involve an enumeration of all possible cases, since there are infinitely many maps. They devised an ingenious procedure to reduce the problem to one which could be solved in a finite length of time. Unfortunately they were not able to solve it by hand because the finite problem still involved too many cases, and they had to use 1200 hours of computer time to complete the proof. In spite of subsequent simplifications of the method, the original proof was never fully checked by other mathematicians. Recently an independent but related proof needing considerably less computer time has been completed by Robertson, Sanders, Seymour, and Thomas. It therefore seems almost certain that the theorem is true, but its proof is still not fully comprehensible.

It is only fair to say that many mathematicians reacted rather negatively to this proof of the four colour theorem. In their view the issue was not *whether* the theorem was true, but *why* it was true (if indeed it was). The computer here acts as an oracle: it tells you the answer, but it is beyond your powers to check its calculations. If mathematics is about understanding, that is *human* understanding, then no satisfactory solution of the problem has yet been found.

Tymoczko put it differently: the proof of the four colour theorem marks a fundamental philosophical shift in mathematics. It makes the truth of at least one theorem an empirical matter, in the sense that we have to rely on evidence from outside our own heads to complete the argument.[2] What of the future? One possibility is that more and more problems will be discovered whose solution can only be obtained by an extensive computer-based search. Many mathematicians fervently hope that this will not happen, but it is entirely plausible.

Goldbach's Conjecture

This famous conjecture, proposed by Goldbach in a letter to Euler in 1742, states that every even number greater than 4 is the sum of two odd primes. Its truth is still unknown, although a large number of similar conjectures have now been proved. The conjecture has been confirmed for all numbers up to 10^{14}, which would be sufficient evidence of its truth for anyone except a mathematician. It has also been proved that it is asymptotically true in the sense that if one lists the exceptions (assuming that there are some), they become less and less frequent as one progresses.[3]

It may turn out that Goldbach's conjecture is similar to the four colour theorem, and that a proof will depend upon a large computer search. Zeilberger even describes the possibility that mathematics might develop into a fully empirical science.

> *I can envisage a paper of c.2100 that reads: We can show in a certain precise sense that the Goldbach conjecture is true with probability larger than 0.99999 and that its complete proof could be determined with a budget of $10 billion.*[4]

There is, however, a worse possibility. Suppose that Goldbach's conjecture is false and that

1324110300685 (a million further digits) 75692837093348572

is the smallest number which cannot be represented in such a way. Suppose also that the shortest way of proving the falsity is by a brute force search. In such a case it is unlikely that the human race will ever discover that the conjecture is false, even if we are allowed to make full use of computers as powerful as we will ever have.

Fermat's Last Theorem

Fermat's last theorem (FLT) is the proposition that it is impossible to find positive integers a, b, c and an integer $m \geq 3$ such that

$$a^m + b^m = c^m.$$

In 1637 Fermat wrote a marginal note in a book claiming that he had found a proof that his equation was insoluble. Nobody now takes his claim seriously, although there is no reason to doubt his sincerity. An enormous amount of work on the problem eventually led to the result that if Fermat's equation does have a solution with $m \geq 3$ then $m \geq 1000000$.

Many editors of mathematical journals received papers claiming to have found proofs of FLT. The one which sticks in my memory came from someone who claimed that the problem was mis-stated. Fermat was supposed to have claimed that there did not exist positive numbers a, b, c such that $a^m + b^m = c^m$ for all $m \geq 3$. There are three problems with this theory. Firstly it is wrong. Secondly this new version is entirely trivial. Thirdly no mathematician cared what Fermat had written, or even whether he had ever existed. The point is rather that what is called FLT is a very interesting and deep problem *whether or not* it was devised by Fermat! Mathematicians use names as *labels*, and regularly attribute theorems to people who would not have understood even the statements, let alone the proofs.

Fermat's problem is not simply an isolated puzzle, of interest only to number theorists. It is part of a subject called the arithmetic of elliptic curves, which has ramifications throughout mathematics. Indeed elliptic curves provide the best current algorithms for factoring large integers, a matter of enormous practical importance in modern cryptography.

In June 1993 Andrew Wiles, a British mathematician working in Princeton, New Jersey, came out of a period of about seven years of near monastic seclusion to give a lecture course on elliptic curves at the Isaac Newton Institute in Cambridge, England. At the end of this he announced that he had solved Fermat's problem! The news appeared in newspapers all over the world, making him an instant celebrity, a unique achievement for a pure mathematician. Wiles acknowledged a serious error in the proof in December 1993, but with the help of Richard Taylor he patched this up within a year and the result was solid. This is one of the hardest mathematical problems solved up to the present date. The proof is beyond the intellectual grasp of most of the human race, and would take about ten years for a particularly gifted 18 year old to understand.

This problem took over three hundred years from its initial formulation to its solution, during which period many partial results and insights were obtained.

When the time was ripe it needed about seven years out of the life of one of the most outstanding mathematicians of the twentieth century to obtain the solution. But at least the result could be grasped in its entirety by a single person of sufficient ability and dedication. Our next example is far worse in this respect.

Finite Simple Groups

A group is a mathematical object containing a number of points (elements) in which multiplication and division are defined, but not addition. Groups are of major importance in mathematics for the description of symmetries, or rotations, of objects. There are 60 rotations of the dodecahedron below (Figure 4.2) which take it back to exactly the same position, including five around the axis shown. Other polyhedra, even those in higher space dimensions, have their own symmetry groups.

Mathematicians have long wanted a complete list of symmetry groups. Among these are some which are regarded as the most fundamental, or 'simple', because they cannot be reduced in size in a certain technical sense. In 1972 David Gorenstein laid out a sixteen point programme for the classification of finite simple groups, and by the end of the decade a worldwide

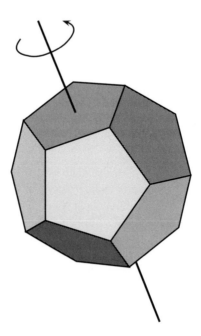

Fig. 4.2 Rotation of a Dodecahedron

collaboration under his leadership had led to the solution of the problem. The final list can be written down in a few lines, and contains a small number of exceptional, or sporadic, groups, the biggest of which is called the Monster. This can be regarded as the rotation group of a polyhedron, but in 196883 dimensions rather than the usual three dimensions of physical space!

---

cyclic groups of prime order
alternating groups on at least five letters
groups of Lie type
26 exceptional groups.

Although the list is short, the complete proof was thousands of pages long, and some crucial aspects were never completed (a mathematician's way of saying that there were serious mistakes in some of the papers). A new project to write out a simplified proof is likely to involve twelve volumes and more than 3000 pages of print.

We have described a theorem whose proof only exists by the collective agreement of a community of scholars. In 1980 none of them understood the entire structure, and each had to trust that the others had done their respective parts thoroughly. Since then the amount and variety of confirming evidence makes it essentially certain that the basic results in the theory are correct, even if individual parts of the proof are faulty. Mathematics has certainly changed since the time of the classical Greeks!

A Practically Insoluble Problem

What lessons can we learn from these examples? It is already the case that understanding the proofs of some theorems takes much of the working lives of the most mathematically able human beings. Extrapolating from these examples, there is no reason to believe that all theorems which are provable 'in principle' are actually within the grasp of sufficiently clever humans. We next give examples of (admittedly not very interesting) statements which are unlikely ever to be proved or disproved.

The first uses the number π, most easily defined as equal to the circumference of a circle of diameter 1. The problem depends upon being able to compute the digits of π successively, but this is in principle straightforward, and the first hundred billion digits have indeed been computed. The first five hundred are given below.

$$\pi \sim 3.14159265358979323846264338327950288419716939937510$$
$$5820974944592307816406286208998628034825342117067 9$$
$$8214808651328230664709384460955058223172535940812 8$$
$$4811174502841027019385211055596446229489549303819 6$$
$$4428810975665933446128475648233786783165271201909 1$$
$$4564856692346034861045432664821339360726024914127 3$$
$$7245870066063155881748815209209628292540917153643 6$$
$$7892590360011330530548820466521384146951941511609 4$$
$$3305727036575959195309218611738193261179310511854 8$$
$$0744623799627495673518857527248912279381830119491 2$$

We will call a number *n untypical* if the *n*th digit of π and the 999 digits following that are all sevens.[5] To find out whether a particular number is *untypical* one carries out a routine calculation which is bound to yield a positive or negative answer within a known length of time.

In spite of the above, the simplest questions about such numbers cannot be answered at present, and may well never be answerable. It is not even known whether there are any untypical numbers. Here are arguments in favour of the two extreme possibilities. If one computes the first one hundred billion digits of π one finds that no number smaller than 10^{11} is untypical. Thus (non-mathematical) induction suggests that there are no untypical numbers. On the other hand, the digits of π satisfy every test for randomness which has been applied to them, and if the digits were indeed random then it could be proved that a sequence of a one thousand sevens must occur somewhere in the sequence of digits. It is an interesting fact that many mathematicians prefer the second argument to the first, in spite of the fact that logically it is even more shaky. It uses non-mathematical induction in that it refers to a finite number of other tests of randomness which imply nothing about the question at hand. Secondly the digits of π are certainly not random: they can be computed by a completely determined procedure in which randomness plays no part.

I cannot refrain from commenting that in my first draft of the above paragraphs I referred to the chain 0123456789 instead of the chain of a thousand sevens. Unfortunately I did not know that it had been proved by Kanada and Takahashi in 1997 that this chain does occur in π. The 0 in its first occurrence is the 17,387,594,880th digit of the decimal expansion of π. There has been great progress in methods of computing the digits of π over the last ten years. However, such developments cannot possibly enable us to decide whether the decimal expansion of π contains a sequence of a thousand consecutive 7s. To demonstrate this, let me make three assumptions. The first is that computational and theoretical progress may one day be so great that it becomes possible to determine a trillion coefficients of the decimal expansion of π every second, with no reduction in speed as one passes along the list of coefficients, however far one has to go. The second is that the only method of proving the existence of the required sequence is by a brute force search. The third is that the first occurrence of the sequence is more or less where it would be for a completely random sequence of digits.

Under these assumptions, none of which is proved of course, we can estimate how long it would take to find the first occurrence of the sequence. I will make no attempt at rigour, and the conclusion can only be regarded as an extremely rough approximation. If we break the expansion of π up into blocks of length 1000, then the chance that a particular block consists entirely of 7s is 1 in 10^{1000}, so one would expect to have to consider something like 10^{1000} blocks in order to find its first occurrence; of course the sequence may not occupy a single block neatly, but this problem may be taken into account. We need not compute every digit of π, but must compute at least one in every block of 1000, because if we leave any block unexamined we may have missed the sequence. Under our standing assumptions we then find that we probably have to compute of order 10^{1000-3} digits and this will take us about 10^{985} seconds. This is vastly longer than the age of the universe, so we had better hope that one of the above assumptions is wrong (if we hope to find the sequence).

A Platonic mathematician would say that either there exists an untypical number or there does not. This view is certainly psychologically comfortable, but it is not necessary to accept it in order to be a mathematician. Intuitionists would only say that such a number existed if they knew one or had a finite procedure which would definitely find one. They would only say that there was no such number if they could derive a contradiction from its existence. If neither was (currently) the case, they would remain silent. They would say that to do otherwise would be to adopt a purely philosophical position which would not increase human knowledge. We will discuss this in more detail in the next chapter.

Warning To make a claim that a mathematical problem will never be solved is perhaps foolhardy. The eminent mathematician Littlewood once wrote that 'the legend that every cipher is breakable is of course absurd, though widespread among people who should know better'. He proceeded to describe an 'unbreakable' code based upon a public coding procedure, a public book of log tables and a private key word of five digits. Fifty years later his code could be broken by standard desktop computer in a few minutes! I hope and believe that I am on safer ground than he was. If I am wrong either computation or mathematics will have advanced beyond the wildest dreams of current mathematicians.

Algorithms

In this section we will discuss some problems which are hard in the sense of computational complexity. Some of these are completely soluble by carrying out a systematic search through all possibilities. However, this method of approach is often completely unrealistic not only for present-day computers but for any computers which could ever be designed. The examples all relate to the behaviour of certain types of algorithm. To make sure that we start from a common position, let me describe an algorithm as a procedure which is applied repeatedly and systematically to an input of a given type. This description is rather forbidding and we start with a simple but famous example.

The Collatz algorithm has as input a single number n. It carries out the following procedures:

If n is even replace n by $n/2$.
If $n \neq 1$ is odd replace n by $3n + 1$.
If $n = 1$ then stop and print YES.

If we start with $n = 9$ then the Collatz algorithm successively yields the values

$$9, 28, 14, 7, 22, 11, 34, 17, 52, 26,$$
$$13, 40, 20, 10, 5, 16, 8, 4, 2, 1$$

so the sequence stops after 19 steps and prints YES.

All algorithms of interest to us have a stopping condition, and when it is satisfied they print an output, which in this case can only be the word YES. For other algorithms there may be several possible different outputs.

An algorithmic solution to a certain kind of problem is an algorithm which is guaranteed to provide the solution to all problems of the specified type. The Collatz problem is whether the Collatz algorithm stops after a finite number of steps, whatever value of n you start from. Surprisingly the answer to this problem is not known, although the algorithm does stop for all n up to 10^{12}. It might seem that one can settle this problem simply by running the algorithm and waiting, and indeed this is true for those values of n for which the algorithm does indeed stop. However, if there exists a value of n for which the sequence is infinite, then this cannot be discovered by use of the algorithm. The fact that it has not stopped after 10^{12} steps says nothing about what might happen after more steps. No solution to the Collatz problem is known, and it is not likely that the situation will change soon.

There are problems which are algorithmically undecidable: there is no systematic way of solving all the problems of the specified type. This is a very strong statement, much stronger than saying that no algorithm has yet been discovered. It only makes sense if one is absolutely precise about what counts as an algorithm, but this has been done in a way which commands general assent. It can then be proved absolutely rigorously that algorithmically undecidable problems exist. We will not discuss this issue further, since it is very technical and has been treated in great detail in several other places.

Let us return to the simplest type of algorithmic problem, one for which it is quite clear that it can be solved in a finite length of time just by testing each one of a large but finite number of potential solutions. Algorithms for such problems are divided into two types, which we will call Fast and Slow.[6] All useful algorithms are Fast, but some Fast algorithms are not fast enough to be useful.

The speed of an algorithm depends upon how one decides to measure the size of the input and also how one defines steps/operations. For most purposes one regards multiplications and additions of large numbers as single operations, and just asks how many are performed. However the amount of time spent transporting data to and from the memory of the computer is also important and one might include such acts as operations.

If we ask for the number of multiplications needed to compute 2^n, that is 2 times itself n times, it seems obvious that the answer is n. However, there is a much better method which involves radically fewer multiplications. Namely we write

$$2 \times 2 = 4$$
$$4 \times 4 = 16$$
$$16 \times 16 = 256$$
$$256 \times 256 = 65536$$
$$65536 \times 65536 = 4294967296$$

which yields 2^{32} in just 5 multiplications. One can actually compute 2^n for general n using far fewer than n multiplications.[7]

Given that a problem may be solved in various different ways, it is obviously desirable to find the most efficient possible way. A problem is said to be (computationally) hard if the number of operations needed to solve it increases extremely rapidly as the size of the problem increases, *for all possible algorithms*. This is clearly difficult to know. It may be that every algorithm currently known for solving a particular problem is Slow, and that nobody believes that a faster algorithm can be found, but that is different from proving that no Fast algorithm can ever be found.

There is one respect in which the idea of thinking of a multiplication as a single operation is misleading. Suppose we have two numbers, one with m digits and the other with n digits. If m and n are large enough then a normal processor cannot multiply them in one step, and they have to be treated as long strings of digits. One possibility is to multiply them using a computer analogue of primary school long multiplication. The number of multiplications and additions of digits is of order $m \times n$. So every algorithm involving sufficiently large numbers is actually much slower than our previous discussion indicated.

The above analysis of algorithms suggests that the only issue in the design of algorithms is to minimize the number of elementary arithmetic operations. This is far from being the case. In all early computers and many current computers there is only one processor, so arithmetic operations do indeed have to be carried out one at a time. However, parallel computers have many processors, so one can carry out multiple operations simultaneously provided one can find an appropriate way of organizing the computation and managing the flow of information between processors.

Suppose one wants to add 512 different numbers (e.g. salaries). The obvious way of doing this takes 511 clock cycles (computers run on a very precise schedule of one operation per clock cycle). If, however, one has an unlimited number of processors, one can add the numbers in pairs in the first clock cycle leaving only 256 numbers to add. In the second clock cycle one can then add the remaining 256 in pairs. Continuing this way the task is finished in 9 clock cycles.

There are many problems in implementing this idea. Firstly most of the processors spend most of their time doing nothing, which is very wasteful. Secondly all of the data has to be moved to the appropriate processors before the computation can start, whereas in the normal algorithm one only needs to bring one item at a time. Thirdly it might not be possible to parallelize some problems at all. Nevertheless the size of many computations in physics is now so large that enormous efforts are being made to find ways of solving the communication and other design problems associated with building large parallel machines. From a purely theoretical point of view the difficulty of an algorithm is now seen to depend on the computer architecture as much as on the problem itself.

How to Handle Hard Problems

Sometimes a problem is extremely hard to solve in the sense that the only known algorithms for solving it are very slow. Two methods for sidestepping this problem have been devised. The first is that one may ask not for the best solution but merely for a good enough solution. Here is an example.

Define $n!$ to be the result of multiplying all the integers $1, 2, 3, 4, \ldots,$ $(n-1), n$ together. To evaluate this we need to perform n multiplications. On the other hand Stirling's formula provides an extremely good approximation to $n!$ which may be computed far more rapidly.[8] It enables one to obtain

$$1000! \sim 4.0238726 \times 10^{2567}$$

with only 10 operations if one regards taking a power as a single operation, or about 30 operations if we use the method already described for computing powers. The following table shows that Stirling's formula is extraordinarily good even for very small numbers:

n	1	2	3	4	5	6
$n!$	1	2	6	24	120	720
Stirling	1.002	2.001	6.001	24.001	120.003	720.009

This illustrates the general fact that if one is prepared to compromise a little on the accuracy or quality of a solution, a problem may become radically easier.

The second method of evading intractable problems is probabilistic. The most famous case of this is finding whether a very large (e.g. hundred-digit) number is a prime. One cannot simply divide the number by all smaller numbers in turn and see if the remainder ever vanishes. The task would take the lifetime of the Universe even for a single thousand-digit prime. In 1980 Michael Rabin devised a probabilistic procedure which solves this problem rapidly but with an extremely small chance of giving the wrong answer. This is now used in commercial encryption systems which transfer money between banks and over the internet. I will not (indeed could not!) describe the procedure, but refer to page 178, where a different and much simpler probabilistic algorithm is described. It marks another step in the transformation of mathematics into an empirical science.

In the last few weeks a Fast (deterministic polynomial) procedure for deciding whether a given number is a prime, without using probability ideas, has been announced by Agrawal et al.[9] The simplicity of this algorithm came as a shock to the community, but it appears to be correct. Such discoveries, and the possible new vistas they open up, are among the things which make it such a joy to be a mathematician. Fortunately (or unfortunately depending on your political views) it does not affect the security of the RSA encryption algorithm. Nor, in practice, is it faster than the existing probabilistic algorithms, but who can tell what future developments might bring?

Notes and References

[1] Hollingdale 1989, p. 148

[2] Tymoczko 1998

[3] Technically the statement is that if $m(n)$ is the number of exceptions less than n then $\lim_{n \to \infty} m(n)/n = 0$.

[4] Zeilberger 1993

[5] I have taken the idea for this example from Gale 1989, but it goes back to Brouwer in the 1920s.

[6] We say that an algorithm is Fast, or polynomial, if for *every* problem of size n the algorithm solves the problem in at most cn^k steps, for some constants c, k which do not depend on n. Both c and k may be of importance for problems of medium size but for very large problems the value of k is usually more significant.

[7] The actual number of multiplications needed is the smallest number greater than or equal to $\log_2(n)$.

[8] This formula, namely

$$n! \sim (2\pi)^{1/2} n^{n+1/2} e^{-n + \frac{1}{12n}}$$

is usually attributed to the eighteenth century Scottish mathematician James Stirling. It was actually discovered by de Moivre, who used it for applications in probability theory.

[9] Agrawal et al. 2002

5
Pure Mathematics

5.1 Introduction

The goal of this chapter is to demolish the myth that mathematics is uniquely free of controversy, and therefore a guaranteed source of objective and eternal knowledge. To be sure, attitudes in the subject generally change very slowly. At any moment there seems to be an overwhelming consensus, provided one excludes a few mavericks. However, this consensus has changed several times over the last two hundred years.

Among the debates within the subject one of the most important concerned its foundations. This was most active in the period between 1900 and 1940. It led to an enormous amount of interesting work in logic and set theory, but not to the intended goal. Indeed the foundations were seen to be in a more unsatisfactory state at the end of the period than they had appeared to be at the start. We describe how this came to pass.

I believe that we are now in the early stages of yet another, computer-based, revolution. Some of my colleagues may disagree, but when one of the lines of investigation into the Riemann hypothesis in number theory involves examining the statistics of millions of numerically computed zeros, something has surely changed. We will discuss this further near the end of the chapter.

Mathematicians themselves rarely have any regard for the historical context of their subject. They attach names to theorems as mere *labels*, without any interest in whether the people named could even have understood the statements of 'their' theorems. Each generation of students is provided with a more streamlined version of the subject, in which the concepts are presented as if no other route was possible. The order in which topics are presented in a lecture course may jump backwards and forwards hundreds of years when compared with the order in which they were discovered, but this is almost never mentioned.

Of course this is defensible: mathematics is a different subject from the history of mathematics. But the result is to leave most mathematics students ignorant of the process by which new mathematics is created. I hope that what follows will help a little to correct this imbalance.

5.2   Origins

The origins of mathematics are shrouded in mystery. One of our earliest sources of information comes from the discovery of hundreds of thousands of clay tablets bearing cuneiform text in Mesopotamia. A few hundred contain material of mathematical interest. From them we glean many interesting but isolated facts about the knowledge of the Babylonians as early as 2000 BC. Among these are their creation of tables of squares and cubes of the numbers up to 30 and their ability to solve quadratic equations. They explained their general procedures using particular numerical examples, since they had no algebraic notation in the modern sense. One of the tablets, dating from about 1600 BC, contains the extremely good approximation

$$1 + \frac{24}{60} + \frac{51}{60^2} + \frac{10}{60^3} \sim 1.4142130$$

(in our notation) to the square root of 2. The tablet called Plimpton 322, dating from before 1600 BC, shows that they had a method for generating Pythagorean triples such as

$$3^2 + 4^2 = 5^2$$

and

$$119^2 + 120^2 = 169^2$$

long before the time of Pythagoras. Such triples were familiar in China and India at a very early date, and there is some evidence for a common origin of this and other mathematical knowledge.

  For many people mathematics means formulating general propositions and proving them by logical arguments from some agreed starting point. In this sense mathematics started in classical Greece, as did so many other aspects of our civilization. After that glorious but brief period centuries were to pass before the subject changed substantially. From about 800 AD Arabic mathematicians started a major development of algebra, arithmetic, trigonometry, and many other areas of mathematics. These percolated slowly through to Europe, and were often described as European inventions until quite recently. During the seventeenth century the focus of development shifted decisively to Europe, where it stayed until some time in the twentieth century. This section describes the historical development of geometry, and the changing philosophical beliefs about its status over the last two hundred years. Later sections discuss logic, set theory and the real number system from a similar point of view.

Greek Mathematics

The main codification of the Greeks' work in geometry was due to Euclid around 300 BC, but Archimedes' importance as an original thinker was certainly much greater. Euclid's *Elements* appeared in thirteen books, with two later additions by other authors. These were preserved during the European dark ages by the Arabs; the first translation available in Europe was that of Adelard in 1120.

The achievement of the Greeks in geometry was revolutionary. They transformed the subject into the first fully rigorous mathematical discipline based upon precisely stated assumptions (called axioms), and proceeded to build a massive intellectual structure using rigorous logical arguments. The method of proof which Euclid used was regarded as the model for all subsequent mathematics for almost two thousand years. Indeed Euclid was still taught in some schools in England in the mid-twentieth century.

We have already encountered the mysterious number π. In the ancient world this was often approximated by 3 or 22/7. Among Archimedes' claims to fame is the first serious attempt to evaluate it accurately. By putting a regular polygon with 96 sides inside the circle, and a similar one outside, he was able to *prove rigorously* that

$$\tfrac{223}{71} < \pi < \tfrac{22}{7},$$

a result which in decimal notation becomes $3.1408 < \pi < 3.1429$. In the third century AD the Chinese mathematician Liu Hui used a polygon with 3072 sides to obtain the more accurate value $\pi \sim 3.14159$, in our notation. I invite readers to obtain a similar approximation to π themselves using hexagons, as in figure 5.1. The experience and the result will surely convince them of the ability of the mathematicians of these ancient times.

It is important to understand what the Greek mathematicians did not do, and why they could not do it. They were greatly hampered by the absence

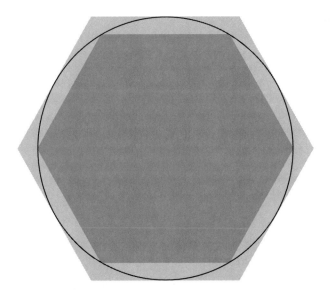

Fig. 5.1 Approximating π Using Hexagons

of a suitable notation for performing numerical calculations. Hindu mathematicians were far ahead of them in this respect. Also algebra, a word of Arabic origin, simply did not exist at that time. In spite of this, Archimedes got closer than anyone else for almost two thousand years to inventing the calculus. He obtained the formula for calculating the area of a sphere by combining several ingenious geometrical constructions. These involved subdividing it into circular strips, summing the areas and using a limiting procedure. This method had been invented by Eudoxus slightly earlier. In our terms it involved being able to sum the series

$$\sin(x) + \sin(2x) + \cdots + \sin(nx)$$

while both describing this equation and proving it by purely geometrical methods![1]

Greek mathematicians also had a more or less complete understanding of procedures for solving quadratic equations. Lacking the algebraic notation to write down the answer, they had to use words instead of formulae to explain how to go about extracting the solution. They grappled seriously with cubic equations and understood without proof that this could not be done using ruler and compass constructions. They devised a number of machines involving sliding pieces of slotted wood (cissoids, conchoids, etc.), which enabled them to solve particular cubic equations. Of course some questions, for example the insolubility of general polynomial equations of fifth degree, were beyond their grasp; even the language for posing these problems did not exist.

It is worth emphasizing that Euclid's approach to geometry did not conform to the Platonic standard. Proposition 11 of Book 11 of *Elements*, for example, starts as follows:

> To draw a straight line perpendicular to a given plane from a given elevated point
>
> *Let A be the given elevated point, and the plane of reference the given plane. It is required to draw from the point A a straight line perpendicular to the plane of reference.*
>
> *Draw any straight line* BC *at random in the plane of reference, and draw* AD *from the point A perpendicular to* BC.
>
> *Then if* AD *is also perpendicular to the plane of reference, then that which is proposed is done. But if not, draw DE from the point D at right angles to* BC *and in the plane of reference, draw AF from A perpendicular to DE, etc.*

Note that the problem is not to prove the existence or uniqueness of such a line, but to describe a sequence of procedures which produce the line, and then to justify it. Euclid's geometry was constrained by his decision to use only ruler and compass constructions. The formulation of the problem and its solution presuppose the existence of a person who does the drawing. Each sentence could have been rewritten to avoid any reference to drawing, as Plato would have demanded, but Euclid did not choose to write in such an artificial style.

The Invention of Algebra

The algebraic notation which we are now taught in school was invented between about 1590 and 1640 in a series of stages, principally by Viète and Descartes, but with several others contributing. Around 1500 Chuquet had written

$$.7.^{2}\,\bar{p}.6.^{1}.\bar{m}\,3$$

to stand for $7x^2 + 6x - 3$. The omission of any symbol for the variable x clearly would have made it difficult to contemplate equations involving several variables. A crucial advance was made by Viète in the 1590s, when he introduced the idea of representing variables by single letters. In his notation

$$B3 \text{ in } A \text{ quad } - D \text{ plano in } A + A \text{ cubo aequator } Z \text{ solido}$$

stood for $3BA^2 - DA + A^3 = Z$. Note the words plano, solido, used to signify when a variable is intended to represent area or volume. Descartes' vital contribution was to remove these words *and also* the requirement that equations should satisfy any corresponding homogeneity condition. Essentially he invented our modern algebraic notation, with the extraordinary power and flexibility it provided.

Viète, Fermat, and Descartes all applied these algebraic methods to the study of geometric problems. In *La Géométrie*, published in 1637, Descartes systematically assigned letters to the lengths of the edges of geometric figures, and then transferred geometric information from the figures into a collection of algebraic equations. By manipulating these equations he obtained the solution of Pappus' locus problem, one of the most famous geometrical problems bequeathed by antiquity. These mathematicians also used their algebraic tools to describe plane curves in terms of what we now call the Cartesian coordinates of points on the curves. These ideas now provide the very language in which mathematics is done.

The Axiomatic Revolution

In spite of Descartes' revolutionary approach to solving geometrical problems, the status of geometry did not change. The basic axioms of Euclid were regarded as being irreducible true statements about the properties of perfect lines, points, and circles in an idealized world. Euclid recognized that one of his axioms, called the parallel postulate, was more complicated than the others. In the fourth century AD Proclus proposed the version

> Given a line and a point not on the line it is possible to draw exactly one line
> through the given point parallel to the line

which mathematicians now call Playfair's axiom with their usual disregard for historical accuracy. We will say that two straight lines are parallel if they do not cross—other definitions are possible, but one has to be absolutely precise before trying to prove anything! Many unsuccessful efforts were made over two thousand years to prove this axiom or replace it by something more natural. Some mathematicians did indeed devise 'proofs', but all such attempts were shown to have flaws.

At the end of the eighteenth century Immanuel Kant described Euclidean geometry as being both a priori (not dependent on external experience) and synthetic (not deducible by unaided logic). He considered that humans have an intrinsic ability to understand geometrical relationships, and this informs our interpretation of the physical world. The fact that we have no choice but to interpret the world in (Euclidean) geometrical terms does not imply that space and time exist in the world itself.

Although Kant is one of the fundamental figures in philosophy, his description of the philosophical status of Euclidean geometry was comprehensively demolished in the nineteenth century. The change of attitude started with work of Bolyai, Lobachevskii, and Gauss. They independently developed the subject now known as hyperbolic geometry. The crucial innovation was that Euclid's parallel axiom was not true in this new geometry. The familiar theorem of Euclidean geometry that the sum of the angles of a triangle is always 180° is replaced by one in which the sum is less than 180°. It was proved that the bigger the triangle the greater the discrepancy in the sum of the angles. Hyperbolic geometry was not merely an aberration with no proper geometrical content. It could be readily interpreted as the appropriate geometry if one is on a certain type of surface, just as spherical geometry was the geometry needed for calculations of distances and angles on the surface of the Earth.

Gauss, by far the most famous of the three, wrote in 1817 that:

> *I am becoming more and more convinced that the necessity of our geometry cannot be proved, at least not by human reason. Perhaps in another life we will be able to attain insight into the nature of space which is now unattainable. Until then we must place geometry not in the same class with arithmetic, which is purely* a priori*, but with mechanics.*

Gauss spent a substantial part of his life directing a geodetic survey of the kingdom of Hannover, inventing an instrument called a heliotrope to make the results considerably more accurate than had previously been possible. (His involvement was commemorated on the German ten mark note, which portrayed him on one side and his heliotrope on the other.) It might be conjectured that he used this as an opportunity to look for deviations of space from a non-Euclidean structure, but this is not plausible: the extreme accuracy of Newton's theory as applied to the planets already implied that any deviations would be far smaller than geodetic measurements could detect. He never published his work on non-Euclidean geometry, partly because of the lack of evidence of its physical relevance, and partly because of a fear of ridicule for adopting such an unfashionable attitude.

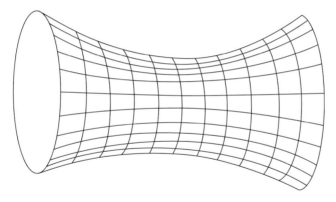

Fig. 5.2 A Curved Surface

In the 1850s Riemann, then a young man, took geometry still further from the familiar Euclidean world. He envisaged geometries of arbitrary space dimensions, and with the curvature of space varying smoothly from point to point. He showed that one could study the geometry of any manifold (curved surface or even curved space), and the analogues of lines, planes, triangles and spheres within it. The idea of *proving* Euclid's parallel axiom was seen to be a chimera, and it was instead understood as the property which distinguished flat space from a huge variety of other equally interesting geometries.

I occasionally get letters from people who object to such ways of evading the parallel axiom as mere sophistry. The latest arrived last week! The authors typically argue that *of course* one can make anything false by changing the meaning of the words involved, but the issue is whether the parallel axiom is true for actual straight lines, and it obviously is. There are two answers to this. One is that since Euclid mathematicians have been interested in proofs at least as much as in truth, and 'elementary' proofs of this particular property have never survived careful scrutiny.

The other problem is easier to explain now than it was in the nineteenth century. What exactly *is* a straight line? The idea seems elementary, but there is an obvious circularity in defining it as the kind of line which can be drawn using a straight ruler. Nor can one define it as the path of a light ray, since we now know that light rays are bent by passing through intense gravitational fields. A straight line cannot be identified *physically* as the shortest route between two points: at cosmological distances the phenomenon of gravitational lensing shows that there may be several different shortest routes between two points. All other attempts to give a definition of straightness turn out, upon examination, to depend upon assumed regularities of the real world. Since Einstein we know that these regularities are only approximate.

The fact that the parallel axiom may not be true *in the real world* is easily understood by assuming that the universe is not in fact infinite in extent, but merely so large that we have no hope of seeing its boundary (indicated

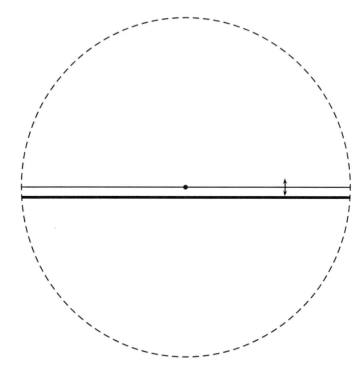

Fig. 5.3 Two Parallel Lines

by a dashed line in the figure). In this case if one turns the 'parallel' line (the thinner one in the figure) about the given point *extremely* slightly, then it will not intersect the other line, because they would have to meet so far away that the universe would not extend that far. This fact would have no impact in normal situations, where lines are either parallel or cross reasonably close to the place of interest, but the parallel axiom would actually be false.

As soon as one accepts that the concept of an infinite straight line only makes sense in an idealized mathematical world, one comes up against the problem that there are several different ways of making the idealization, and none has an obvious claim to being the *right* one. Such considerations gradually led mathematicians away from the idea that geometry was the study of physical space. When David Hilbert wrote *Grundlagen der Geometrie* in 1899, he regarded geometry as a purely axiomatic subject, to be developed by the application of nothing beyond pure logic starting from precisely stated axioms. They had become rules of the game, just like the rules of chess, so it made no sense to ask if they were true. Hilbert also proved that Euclid's axioms were consistent, in the sense that any contradiction derived from them would imply the inconsistency of ordinary arithmetic.

Riemann's approach to geometry turned out to be of crucial importance for Einstein, since it provided exactly the tools he needed to develop his special and general theories of relativity. After Einstein the survival of Euclidean geometry relied upon the fact that we are normally interested only in objects moving at speeds much less than the speed of light in gravitational fields which are fairly weak.

Unfortunately the new reliance on axiom systems caused many pure mathematicians to isolate themselves from other scientists, and to elevate the formal aspect of the subject to a status which it had never previously had. It took Gödel to show the ultimate limits of this approach in 1931, and the development of computers late in the twentieth century to bring some pure mathematicians back to a more empirical way of looking at the relationship between mathematics and the external world.

Projective Geometry

Projective geometry provides an excellent case study of the relationship between axiom systems and their interpretation. The subject studies those properties of straight lines and points which do not involve any mention of the sizes of angles or lengths of lines. It goes back to classical times, one of the most famous theorems, due to Pappus of Alexandria in the fourth century AD, being stated as follows (all lines are assumed to be straight):

> Let L and M be any two lines (the two thicker ones in figure 5.4). Let a, b, c be three points on L and let d, e, f be three points on M. Join these points by lines as shown in the diagram and consider the three points p, q, r. Then these points must also lie on a line (labelled N in the diagram and drawn with dashes).

Projective geometry is of great importance in the mathematics of perspective. It was developed by the Florentine architect Brunelleschi during the Renaissance, but its axiomatic structure was only fully clarified in the nineteenth century. There are several axioms needed for a full description but I only mention the following:

> There are two primitive concepts called line and point.

> There is a relationship between lines and points which can be equivalently stated as 'the line l passes through the point p' or 'the point p lies on the line l'.

> There is a unique line passing through any two points.

> There is a unique point lying on both of any two lines.

The point I wish to emphasize is that the status of lines and points in these and the other axioms is exactly the same. Hence, given any theorem in the subject, one

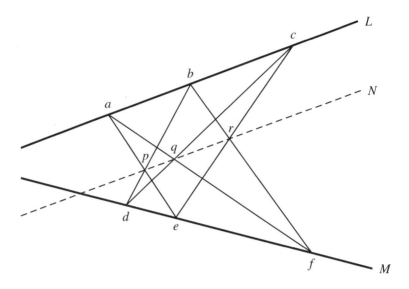

Fig. 5.4 Pappus' Theorem

may interchange the words 'line' and 'point' to obtain another theorem which is necessarily true, because the proof is exactly the same. This phenomenon is called duality. To illustrate it I describe the dual of Pappus' theorem. (It happens that this particular dual theorem is the same as the original, but this requires one to relabel the diagram suitably.)

> *Let L and M be any two points. Let a, b, c be three lines passing through L and let d, e, f be three lines passing through M. Consider the three (dashed) lines p, q, r constructed as shown in the diagram. These lines must all intersect at a single point N.*

Consider two mathematicians who only communicate by email and who only write about projective geometry. Suppose one of them tells the other a series of theorems, but the other does not know which of the words 'point' and 'line' refer to points and lines respectively. Then the two could have completely different pictures in their minds, although they would agree about the truth of every theorem. They would indeed have no reason to suspect that their mental images were totally different.

The easiest conclusion to draw from the above is that the mental images of the two mathematicians are not a part of the mathematics itself: they are no more than psychological aids which humans seem to need when doing mathematics. There is however, a quite different interpretation, which I introduce by an analogy with the manufacture of paper. Although its main application nowadays is to keeping written records, it has always had many other uses,

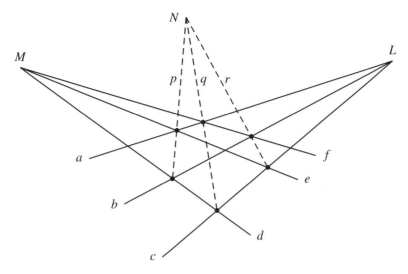

Fig. 5.5 Dual of Pappus' Theorem

from wrapping presents and insulating walls to origami. One might say that paper has little interest (except to manufacturers) until one thinks of using it in some particular way. Similarly theorems have little interest (except to formalists) until one passes from their formal statements to some interpretation, often a geometrical one. This is why mathematicians try to find the idea behind a proof, and feel that they are missing the whole point if they can do no more than check the validity of each line. The possibility of a theorem having several different interpretations is extra richness as far as mathematicians are concerned, and not evidence of human inability to grasp the 'true theorem'. Human mathematics involves the continuous interplay between formal theorems and interpretations. Sometimes one is more dominant and sometimes the other, but neither is dispensable.

5.3 The Search for Foundations

The formalization of logic was started by the classical Greeks, but the advanced technical development of the subject only got started towards the end of the nineteenth century with the work of Frege. By an historical accident Bertrand Russell, destined to become one of the most famous intellectuals of the twentieth century, was influenced more by Cantor and Peano than by Frege. He developed their ideas into a monumental three volume work *Principia Mathematica* written jointly with A. N. Whitehead between 1910–13. He may be regarded as the founder of the logicist approach to the foundations of mathematics, described below.

The theory of sets[2] is considered by some mathematicians to be so fundamental that it is surprising that it was not developed as a subject in its own right until work of Dedekind and Cantor in the 1870s. Cantor showed how to compare the relative size of two infinite sets. The idea developed that logic and set theory were self-evidently valid in a way which the rest of mathematics was not, and that all of mathematics should be derived from them by a process of formal deduction and construction. Frege devoted much of his intellectual life during the last quarter of the nineteenth century to this project, transforming the technical status of logic.

Unfortunately Russell was to find a serious and elementary flaw in the work of Cantor and Frege in 1902. Its precise nature is not central to our discussion, but it went as follows. Let R denote

the set of all sets which are not elements of themselves.

He considered the two cases (i) R is not an element of itself (which you will certainly believe if you consider that the idea that a set might be a member of itself is absurd); in this case R has the property referred to in the definition of elements of R, so R is an element of itself; (ii) R is an element of itself; in this case R does not have the property referred to in the definition of elements of R, so R is not an element of itself.

Russell's paradox is not important in itself, since nobody had any interest in this very peculiar set. The problem was rather that it was not clear what *principle* one could use to eliminate other sets whose paradoxical nature might be far from obvious. The paradox shook Frege's confidence in the transparency of the notion of sets and in the validity of his life's work, since he could see no *systematic* way of avoiding such paradoxes. It led Russell to his theory of types, which imposed technical limitations on the kind of property which could be considered to define a set. Others generally considered that this was at best a clumsy way of resolving the paradoxes of set theory, and a better solution came with the development of the Zermelo–Fraenkel theory of sets in 1908 and 1922. The ZF set theory also avoided the Russell paradox by imposing limits on what kinds of property specify sets, but in a less clumsy way than Russell had proposed. These limits are nevertheless still artificial in the sense that the avoidance of obvious inconsistency is the only reason for their presence. Although the ZF system has withstood the test of time, nobody *knows* whether other paradoxes may turn up and force further fundamental changes in set theory. Nor is this mere pedantry: the study of the foundations of mathematics is full of theories which turned out to be inconsistent, and confident assertions which turned out to be quite wrong.

We now turn to David Hilbert, who dominated the world of mathematics during the early years of the twentieth century. He made the Mathematics Department at Göttingen pre-eminent in Europe, with streams of famous visitors. Among his contributions was the proposal of an entirely new way of laying secure foundations for mathematics. He was not willing to accept the ZF system as truly foundational because of the lack of any proof of its consistency. Nor did Hilbert believe that infinite sets or any other infinite entities actually existed in themselves. In 1930 he wrote:

> *As far as the concept 'infinite' is concerned, we must be clear to ourselves that 'infinite' has no intuitive meaning and that without more detailed investigation it has absolutely no sense. For everywhere there are only finite things . . . And although there are in reality often cases of very large numbers (for instance, the distance of the stars in kilometres, or the number of essentially different games of chess) nevertheless endlessness or infinity, because it is the negation of a condition which prevails everywhere, is a gigantic abstraction.*[3]

In spite of the above, Hilbert was not prepared to abandon any part of classical mathematics, as Kronecker and later Brouwer and Weyl considered necessary. His belief was that by focusing on the syntax of mathematics, that is the formal rules for manipulating mathematical symbols, it would be possible to prove that mathematics was consistent and complete. He was not suggesting, as some thought, that mathematics was a meaningless formal game played by manipulating strings of symbols, but rather that its consistency and completeness could be established by his formalist programme. Once this had been achieved mathematicians would be able to relax in the knowledge that they would never again be caught out as Frege had been.

I should, perhaps, explain the words consistency and completeness. By consistency Hilbert meant that it should not be possible to prove a contradiction within the formal system constructed. Completeness is more interesting. The idea is that if one takes a definite statement within the system, there must always exist *within the system itself* either a proof of the statement or a proof of its incorrectness. Now for Hilbert a proof was merely a string of symbols produced and manipulated according to certain rules. Every such string is of finite length and all possible strings can be listed in lexicographic order. So completeness requires that if one runs through the list of all correctly formed chains of symbols, one will eventually find either a proof of any statement or a proof of its incorrectness.

Hilbert's challenge was taken up by some of his junior colleagues, particularly Paul Bernays, with whom he eventually published *Foundations of Mathematics* in 1934. Much valuable mathematics was produced, and Hilbert repeatedly made his confidence in its ultimate success clear. On his retirement in 1930 he was made an honorary citizen of his native city of Königsberg. The end of his acceptance speech contained the following:[4]

> *For the mathematician there are no unknowable facts, nor, in my opinion, for any part of natural science. The real reason why Comte was unable to find an unsolvable problem is, in my opinion, that there are absolutely no unsolvable problems. Instead of foolish claims about unknowability, our credo claims:*
>
> *We must know. We shall know.*

In that very year the whole programme was dealt a fatal blow by Kurt Gödel. Gödel's first theorem, published in 1931, proved that it was not possible to achieve the goal—all attempts were bound to fail *whatever formal system was used*. In very rough terms Gödel proved that any formal system of sufficient complexity to capture the behaviour of the numbers must be limited in the sense that there will be undecidable statements. These are statements which one can neither prove nor disprove *if one only uses rules within the formal system*. His second major theorem was a proof that the consistency of any such formal system is impossible to prove within the system. His results were highly technical but mathematically decisive, and they came as a bombshell to the community. Eventually they led to the acceptance that there was no way of providing the unassailable foundations which mathematics was considered to need and deserve.

Controversy arose when people (including Gödel!) started to claim that Gödel's discoveries had implications concerning human ability to transcend formal methods. The argument is as follows. Consider the statement that every number has some particular property. If we can check this systematically for $1, 2, 3, \ldots$ then either there is a counter-example or there is not. If a counter-example exists then that fact is revealed within the formal system by checking that the relevant test fails. So if there is no formal proof that the result is false, then no counter-example exists. *We* can conclude that the hypothesis is true, whether or not a proof exists within the formal system. So human beings can transcend any formal system, and hence any computing machine.

This argument assumes that it makes sense to contemplate the result of testing *all* numbers. Equivalently it assumes that there is a matter of fact about the statement, whether or not anybody could ever determine it. Gödel himself was a Platonist and did indeed come to such conclusions. Penrose goes much further than this in *Shadows of the Mind* when he says:

> It will be part of my purpose here to try to convince the reader that Gödel's theorems indeed show (that human intuition and insight cannot be reduced to any set of rules), and provides the foundation of my argument that there must be more to human thinking than can ever be achieved by a computer, in the sense we understand the term 'computer' today.

But Gödel's theorems are about formal systems and make no reference to human abilities. There is much controversy about whether one can legitimately draw such conclusions from his theorems. Penrose's argument is a modification of that of John Lucas dating to the 1960s, and has been heavily criticized on both occasions. Penrose has responded vigorously to these criticisms.

At the other extreme is the view that arithmetic is a human construction within which Gödel demonstrated that there exist meaningful statements *which have no truth value*. There have been so many arguments about this issue that it would be impossible to list them. Michael Dummett has argued that human insight 'is not an ultimate guarantee of consistency, nor the product of a special faculty of acquiring mathematical understanding. It is merely an idea in an embryonic form, before we have succeeded in the task of bringing it to birth in a fully explicit form'.[5] Formalization is the best way we have of making ideas explicit and communicating them. The fact that Gödel proved that it has limitations does not imply that some better method is waiting to be discovered.

5.4 Against Foundations

In this section I describe a number of arguments which have been used with increasing force and confidence since about 1950 to undermine the claim that mathematics needs foundations, whether these are based on set theory, logic, formalism, or anything else. These arguments focus on the way in which mathematics is created rather than the description of the final product. They also have considerable philosophical support, surveyed in Tymoczko's recent anthology *New Directions in the Philosophy of Mathematics.*

A positive case for regarding set theory as being fundamental is that it seems to be possible to reformulate the basic notions of almost any mathematical theory in such terms. The benefits of this are that one has a single theoretical structure within which one may examine the correctness of mathematical proofs. Unfortunately the reformulations are often much less easy to understand than the concepts which they supposedly explain. This does not in itself contradict the possibility that the set-theoretic version is more fundamental. For almost everyone now accepts that the atomic theory of matter is more fundamental than what preceded it, even though the properties of atoms are far more difficult to grasp than the properties of bulk matter. But there is a vital difference. Since its introduction atomic theory has had a profound and increasing practical impact on vast areas of physical science. On the other hand even now hardly any of the deepest theorems in mathematics have depended upon the use of formal set theory, or formal logic.

I will digress to express my frustration about the numerous philosophers who write about mathematics when they obviously know very little of it except for formal logic and set theory. These subjects are not important for mathematicians, the great majority of whom have never taken a course in formal logic and would not be able to write down the Zermelo–Fraenkel axioms of set theory. Some of these philosophers like to bolster their arguments by re-expressing them in formal logic, subjecting their readers to the need to struggle through the equations to work out what they actually mean. I have yet to see a case in which this conveys any idea which could not be expressed just as well in ordinary language.

I should not leave you with the impression that the debate about the status of set theory has been forgotten by all pure mathematicians. In 1971 Paul Cohen wrote the following:

> *Historically, mathematics does not seem to enjoy tolerating undecidable pro-*
> *positions. It may elevate such a proposition to the status of an axiom, and*
> *through repeated exposure it may become quite widely accepted. This is more*
> *or less the case with the Axiom of Choice. I would characterize this tend-*
> *ency quite simply as opportunism. It is of course an impersonal and quite*

> *constructive opportunism. Nevertheless, the feeling that mathematics is a*
> *worthwhile and relevant activity should not completely erase in our minds an*
> *honest appreciation of the problems which beset us.*[6]

A problem with most approaches to the foundations of mathematics is that they
have no relationship with the way in which mathematics is created. There is
an enormous amount of activity in the subject, and providing formal proofs of
theorems which are already well understood comes at the bottom of the mathematical agenda. Indeed a wide range of outstanding mathematicians including
Hardy, Lakatos, Polya, and Thurston emphasize that it involves imagination,
analogy, experimentation, and a variety of other skills in essential ways. It is also
significant that logically incorrect arguments have frequently led to important
new insights. A famous, but by no means uncommon, instance of this occurred
when Euler proved that

$$1 + \tfrac{1}{4} + \tfrac{1}{9} + \tfrac{1}{16} + \cdots = \tfrac{\pi^2}{6}$$

by a method which even he did not regard as justified. But having then checked
the purported answer to several decimal places he convinced everybody that the
result was correct. Only years later did he find a more acceptable proof. From
a formal point of view everything except the very last stage of this process is
essentially meaningless.

There is abundant historical evidence against mathematics being a subject
based on logical deductions from explicit and precise initial premises. The main
properties of the trigonometric functions were established before the end of the
fifteenth century on the basis of a realist understanding of Euclidean geometry.
The development of calculus by Newton and Leibniz in the seventeenth century
preceded the rigorous definition of the real number system by two centuries.
Cauchy's fundamental study of the theory of functions of a complex variable
early in the nineteenth century preceded the rigorous definition of complex
numbers by 'only' fifty years. I could go on, but it hardly seems necessary.

The last twenty five years have seen an increasing use of computer based
methods for investigating mathematical problems. I myself have written several research papers in which my discovery of a certain phenomenon arose
from numerical calculations. The final result of one such piece of work was
an entirely conventional piece of pure mathematics in which the proofs made
no use of computation at any stage. I would reject any suggestion that only
this final product was actually mathematics, since for me, at the time, the two
aspects of the problem were totally intertwined. Combining empirical methods
with traditional proofs, with the empirical aspect leading the way, is becoming
increasingly common even among pure mathematicians.

Should we conclude from all this that what mathematicians spend their
time doing is not mathematics? And should we declare that the way in which
mathematicians create and understand mathematics has nothing to do with the
philosophical status of the subject itself? Since formal proofs of most substantial
theorems have never been produced, mathematicians cannot be relying on them

to justify their belief in the truth of their theorems. David Ruelle put it the following way:

> *Human mathematics consists in fact in talking about formal proofs, and not actually performing them. One argues quite convincingly that certain formal texts exist, and it would in fact not be impossible to write them down. But it is not done: it would be hard work, and useless because the human brain is not good at checking that a formal text is error-free. Human mathematics is a sort of dance around an unwritten formal text, which if written would be unreadable.*[7]

Some people claim that we can be confident that formal proofs of all genuine mathematical theorems could be produced. If one then asks for the basis for this confidence, one discovers that it is no more than the intuition of practitioners in the field. To focus exclusively on the formal aspects of mathematics is to ignore the essential content of the subject, which consists of *ideas*. Published research papers are usually written in a rather forbidding and unmotivated style, and because of this are extremely difficult even for mathematicians to understand. Indeed much of our understanding comes during discussions at a blackboard. The fact that the nature of mathematical ideas is very difficult to examine because of a lack of written evidence does not justify claiming that something else is the essence of the subject. According to André Weil formalism is rather like hygiene: it is necessary for one to live a healthy life, but it is not what life is about. One needs to have experts in logic and set theory, as in hygiene, but their chosen subject is not the basis on which everything else is built. Formal logic is much better thought of as a mathematical discipline in its own right, no more or less fundamental than any other part of pure mathematics.

I do not claim any originality for the above ideas, which have been expressed by many people. For example Lakatos wrote:

> *[The logicists and meta-mathematicians] both fall back on the same subjective psychologism which they once attacked. But why on earth have 'ultimate' tests, or 'final' authorities? Why foundations, if they are admittedly subjective? Why not honestly admit mathematical fallibility . . .*[8]

He quoted von Neumann, Quine, Church, Weyl, and others as accepting that mathematics should be regarded as a semi-empirical science, a position very different from the popular perception of mathematics even today.

An illuminating view of the nature of proof in mathematics was given by Richard Feynman in his book *The Character of Physical Law*. He wrote:

> *So the first thing we have to accept is that even in mathematics you can start in different places. If all these various theorems are interconnected by reasoning there is no real way to say 'These are the most fundamental axioms', because if you were told something different instead you could also run the reasoning the other way. It is like a bridge with lots of members, and it is over-connected; if pieces have dropped out you can reconnect it another way.*

We conclude that mathematics is a human activity with an ineradicable possibility of error. This has been much reduced by the efforts which generations of mathematicians have put into achieving consistency between the many different approaches to the subject. One should imagine mathematics not as a tree in which everything is fed by the roots of logic and set theory, but rather as a web in which every part strengthens every other part.[9]

There is another type of support for the idea that a web is a better analogy for the structure of mathematics than a tree. If mathematicians were only interested in the truth of theorems then as soon as one sound proof of a theorem was known they would move on, never to return. In reality, however, they constantly return with new proofs of familiar theorems, each throwing new light on its connections with other parts of the subject. On the web analogy this is entirely comprehensible, in that every new proof forms new connections and reinforces the structure. Ideally the subject should be so interconnected that a large number of links could be removed without compromising its integrity.

Empiricism in Mathematics

Yet another possibility is to adopt an empiricist point of view towards mathematics.[10] Donald Gillies has argued that it is not profitable to discuss whether mathematics *as a whole* is an empirical or a metaphysical subject. He divides the statements of science and mathematics into four levels. The bottom one consists of those which can be decided by direct observation, while the next two involve theories which are to some degree testable. The top level consists of metaphysical statements, which are too far from observation to be confirmed or refuted even indirectly. He applies this classification to put certain highly infinite sets and numbers into the metaphysical category. Some topics, such as the real number system, are well supported by their use in scientific contexts, while others, such as the theory of large cardinals, are not. To paraphrase his argument:

> *Higher cardinal numbers have a use within the language game of Cantor's set theory. This activity may have few participants, but it is nevertheless a perfectly definite social activity carried on in accordance with clear and explicit rules. On the other hand statements about higher cardinals have as yet no truth value because there is no application in physics which would give them a reference within the material world. In such metaphysical parts of mathematics one may prove certain theorems, but that is a different matter from attributing truth to the theorems.*[11]

In Chapter 3 I suggested that one should also adopt a nuanced attitude towards the status of whole numbers. Small numbers have strong empirical support but huge numbers do not, and only exist after assenting to Peano's axioms. I therefore consider that huge numbers have only metaphysical status. I have argued in a recent article that this does not prevent one using the real number system in the normal way.[12] From the empirical point of view extremely small real numbers have the same questionable status as extremely big ones. But this is perfectly OK: physicists know that extremely small quantities, for example lengths far smaller than the Planck length, have no physical meaning anyway, so this philosophy of mathematics matches the nature of physics perfectly.

From Babbage to Turing

If one had to sum up the life of Charles Babbage in two words they would have to be 'frustrated genius'. Born in 1791, he became a Fellow of the Royal Society in 1816 and Lucasian Professor of Mathematics at Cambridge in 1828. His early work on mechanical computing engines met with acclaim, and over a period of time he was given seventeen thousand pounds in Government grants to construct a machine which would compute mathematical tables automatically. This would bring to an end the many errors which affected all tables produced by hand.

Babbage devoted most of his life to this project. A part of his first difference engine, dated 1832, is housed in the Science Museum, London, but it was never completed because he fell out with his toolmaker, Joseph Clement, the following year. Shortly after that Babbage discovered 'a principle of an entirely new order'. He abandoned his difference engine and started on the design of a much more ambitious 'analytical engine', supported by Ada Lovelace. Unfortunately the Government withdrew its support in 1842, at the express order of the Prime Minister, Sir Robert Peel. From that point on, Babbage had no chance of ever building the engine, which would have been the size of a locomotive. He continued to work on the project as he grew older, but became increasingly disappointed and embittered.

The analytical engine was (i.e. would have been) the first general purpose programmable computer. It was controlled by a pile of punched metal cards, which were read by it one at a time and then acted on. When compared with the Jacquard loom of about 1800, it had a crucial innovation. In certain situations the progression through the cards could be reversed, so that a group of cards could be read again and again. In modern computing terms the engine could implement iterative loops, and even nested loops. This resulted in a dramatic reduction in the number of cards needed, but also a change in the character of what could be calculated using the engine. The design even allowed the engine to print out its results onto paper ready for binding. Figure 5.6 shows a small part of it, measuring about a metre across, as it was at his death in 1871.

Ada Lovelace did not simply provide moral support to Babbage. She made major intellectual contributions, in an era when this was more or less unheard of for a woman. Born in 1815, she was the daughter of the poet, Lord Byron, who abandoned her and her mother a month later. Her mother, who also had mathematical talents, ensured that she had a thorough mathematical education. After she met Babbage in 1833, she became engrossed in his project, and eventually published a book in 1843, describing the operation of the analytical engine. She modestly called the book a translation of an article of Menabrea on Baggage's engine, with notes by the translator, but her notes were twice as long

Fig. 5.6 Part of Babbage's Analytical Engine
By permission of Science Museum, London/Science and Society Picture Library

as the original article, and better informed. She emphasized how much more advanced the analytical engine was than the earlier difference engine, referring particularly to its use of 'cycles' (iterative loops). She included what is surely the first ever computer program, which used the engine to compute Bernoulli numbers. Her description sets up the problem mathematically, specifies the intermediate variables used, and lists the elementary operations $(+, -, \times, \div)$ to be carried out on those variables. This book was to mark the high point of her career. Shortly after completing it she became ill, and died of cancer in 1852.

After Babbage's death, no further development of comparable scope occurred until the middle of the twentieth century. Eventually the development of electronics opened up an entirely different way of building computers, and the Second World War made the expense of developing the technology less important.

The starting point for the new initiative came in 1936, when Alan Turing invented a mathematical idea which is now called a universal Turing machine, and which is frequently considered to encapsulate what is meant by computation within a formal setting. Turing was one of the key people in the (re-)invention of computers in the 1940s, and was the mastermind behind the now famous code breakers of the German Enigma codes in Bletchley Park during the Second World War. He was able to show that there exist problems which even a universal Turing machine cannot solve. This provided the computational counterpart to Gödel's work in logic.

In a famous paper written in 1950, Turing discussed whether computers would ever be able to think.[13] He considered this question too vague for meaningful discussion and proposed a sharper version. Namely would a computer ever be able to conduct a conversation (by letter, say) so well that nobody could distinguish the computer from another person. He expressed the belief that this would happen within about fifty years. His test has become famous, but many people consider that it is not an appropriate way of measuring the ability to think. Gödel considered that Turing had made a serious philosophical error in believing that computers might one day be able to indulge in genuine thought, as opposed to mere simulation of thought.[14] Turing was of course well aware of Gödel's work, and explained in his article why he did not consider it provided any barrier to his conjecture.

The goal of this section is not simply to give yet another account of Turing machines, although we have to start with that. It is to point out some difficulties relating to the widespread belief that they describe perfectly what is *meant* by computation. There are many different ways of describing Turing machines, and we will follow the usual approach, in spite of the fact that it looks extremely dated by the standards of modern computer technology.

One starts with an infinitely long tape, supposed to be laid in a straight line on the floor. The tape consists of a long series of cells, each of which can contain any of the symbols on a computer keyboard. It makes no difference

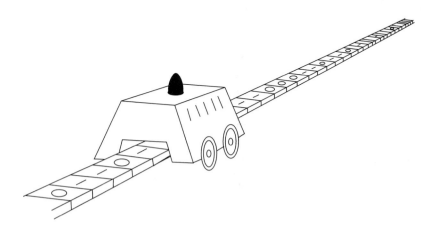

Fig. 5.7 A Turing Machine

to the discussion if we impose the further restriction that each cell can only contain the symbol 0 or 1, since one can group cells together eight at a time and code any keyboard symbol by a sequence such as 00111010. The tape serves as the memory of the computer. Initially a finite number of the cells contain the problem/program, but all the rest contain only the symbol 0. In addition to the tape the computer has a processor, which consists of a machine which can move along the tape one step at a time.

The processor has a finite number of internal states and can view the cell it is currently at. It then makes a decision about whether to alter the symbol in the current cell and whether to move one step to the left or right. Among the rules of the machine is one which tells it conditions under which it should stop and print out its answer to the problem. A Turing machine is said to be universal if its internal rules for making decisions are sufficiently rich. We need not specify this more precisely, other than to say that any current computer running a typical high level programming language such as C^{++} would be a universal Turing machine (UTM) *if it had an infinite memory*. UTMs play a key role in investigating the fundamental limits on what can be computed in an ideal world in which there are no constraints on the size of the memory or on the time taken for the computation. It has been found that there exist problems which definitely cannot be solved by such a machine, and hence that there are limits to what can be proved within formal mathematics.

A more precise statement of the problem is as follows. It is relatively straightforward to test whether a program intended to be run on a particular Turing machine is grammatical, that is whether it makes sense. Some grammatically correct programs will run for a length of time and then print out a result and halt. Others may simply run for ever, because they have no instruction to halt, because they get into a repetitive loop, or because what they are trying to do gets more and more complicated, occupying ever more of the memory tape. Turing discovered that there is no *systematic procedure* for examining a program and deciding whether it will halt or not. This is called the Halting Problem: there are programs which would in fact run for ever, but it is not possible to identify them in a systematic manner. This deep fact implies Gödel's incompleteness theorem. The link is the fact that proofs of theorems can be carried out by a program which runs through all conceivable arguments, checking whether any of them is in fact a formal proof of the required theorem.

The theory of Turing machines has two aspects, which should be distinguished. They provide a context within which one can discuss the existence of formal proofs of mathematical theorems. If a theorem cannot be proved using a Turing machine with an infinite memory, it certainly cannot be proved in a finite context. Whether or not such machines can be built in the real world is not a relevant issue. This aspect of the theory of Turing machines has been an extremely fruitful source of new mathematics since 1936.

The Church–Turing thesis is a more controversial matter. One form states that any operations which can be performed by a computer following precise instructions can also be performed by a universal Turing machine. However, the words do not mean what someone reading them in the year 2000 might imagine! In 1936 a computer was a human being employed to carry out routine calculations by hand. The thesis emphasizes that the person was not allowed to use insight or intuition. Turing machines were concepts: their implementation by electronic hardware was several years in the future. Turing and Church were, however, well aware that Turing's ideas were relevant to the capacities of possible future computing machines. In his review of Turing's 1936 paper Church wrote the following:

> *(Turing) proposes as a criterion that an infinite sequence of digits 0 and 1 be 'computable' (if) it shall be possible to devise a computing machine, occupying a finite space and with working parts of finite size, which will write down the sequence to any desired number of terms if allowed to run for a sufficiently long time. As a matter of convenience, certain further restrictions are imposed on the machine, but these are of such a nature as obviously to cause no loss of generality—in particular, a human calculator, provided with pencil and paper and explicit instructions, can be regarded as a kind of Turing machine.*[15]

There is a strong, or physical, form of the Church-Turing thesis which goes far beyond anything written by Church or Turing. It states that anything which can be computed by a physical computing machine with any conceivable internal architecture in any possible physical world can also be computed by a universal Turing machine (UTM). It is now known that this physical form of the thesis is *simply wrong*!

We start by pointing out that real computers do not have infinite memories, so UTMs are idealizations. They are, moreover, idealizations which are in conflict with the laws of physics in several ways. We mention just two. If the universe is finite then a UTM cannot be built because there will not be enough materials to build it. If the universe is infinite and each memory cell had the same positive mass then a UTM would have infinite gravitational self-energy, and would therefore immediately collapse into a black hole. If one is only concerned with what a Turing machine *can* do then it does not need an infinite memory, because any particular computation which can be completed will only use a finite amount of memory. However, each of the above problems has a finite version, leading to the conclusion that Turing machines with sufficiently large finite memories cannot be built in our universe. In addition any long enough computation could not actually be carried out because the universe will not last long enough in its present form. It is easy to write down computations of this type.

These objections are typically ignored on the grounds that UTMs are perfectly plausible *in principle*, and for a thought experiment this is sufficient. However, much more powerful types of computer are equally possible *as thought experiments*, and there is a growing literature on how to construct them. One of the more exotic possibilities depends upon the fact that time is not absolute in general relativity. It may be possible to shoot a computer into an exotic space-time singularity and observe it *actually carrying out* an infinite number of computations within what is, *for the observer*, a finite length of time.[16]

Another possibility is to consider an infinite hierarchy of mechanical calculating machines, each of which is smaller and faster than the one before. The operator gives a task to the first and biggest in the chain, which then passes it on to the smaller machines according to certain carefully specified rules. One may divide up certain infinite sets of computations between the various machines in such a way that they are able to complete *all of them* in a finite length of time. I have recently given a detailed specification of such a machine and how to build it *in a continuous Newtonian universe*.[17] This is an imaginary world obeying Newton's laws but with no atoms, so that matter may be subdivided indefinitely. Such a machine could 'prove' Fermat's last theorem not by finding a finite chain of arguments, as Wiles did, but by the brute force testing of all potential cases. If any of the machines finds a counter-example to the statement being tested, this fact is reported back through the hierarchy to the first machine. It may not be possible to report back the *value* of the counter-example, because it may be too large to be stored in the memory of the first machine. If no report is received within a certain finite length of time, then no counter-example exists. The collection of machines would, in effect, act as an oracle. Of course these machines are impossible to build in *our* universe, but so are sufficiently large Turing machines.

There are other types of idealized computing machine which are not equivalent to that of Turing, but which are of substantial interest. One of them allows each memory site on the tape to contain a real number rather than one of only a finite number of symbols. Of course this is not possible in fact, but neither is an infinite tape. The issue is which definition is more interesting, and this depends upon what one wants to do with it. Since scientific computation is heavily dependent on manipulating real numbers, there is a good case for studying this new type of machine.[18] This perhaps resolves a complaint of von Neumann about the lack of relationship between Turing machines and the requirements of mathematical analysis, 'the technically most successful and best-elaborated part of mathematics'.

There are computers operating within our own physical world which are not Turing machines. We call them analogue computers. They simulate the world in a non-discrete manner, and were quite popular in the 1950s. They certainly do not fit into Turing's framework and there are claims that they can go beyond the Turing limit of computation. Of course one could say that they are not really computers, but this line of defence turns the strong Church–Turing thesis into a tautology. There is an active debate about whether analogue computers can achieve anything which a sufficiently lengthy digital computation could not.

There is another kind of computer which goes beyond Turing machines in a much more radical manner: the very recently invented quantum computers. It appears to be possible, at least in principle, to construct computers in which the fundamental processing units operate on quantum mechanical principles. Such computers may one day be able to perform certain tasks which are far beyond the scope of traditional machines, and may allow the rapid deciphering of the so called 'unbreakable codes' based on the use of very large prime numbers. If this idea can be implemented at a practical level then it will create yet another computer revolution. At the time of writing (January 2002) it has just been announced in Nature that a quantum computer has succeeded in factorizing the number 15. This may be regarded either as laughable or as a proof that the fundamental concepts behind quantum computation are correct, depending upon one's attitude towards blue skies research. Clearly there is a long way to go before practical quantum computers start being sold. Even if the technical difficulties cannot be surmounted, the appearance of the idea already establishes that computers need not be restricted to the use of classical logic. In particular the view that physical computers and universal Turing machines are effectively the same thing is no longer tenable.

Finite Computing Machines

The abstract theory of Turing machines disregards a crucial factor in all real computers. A program which has to run for a huge length of time in order to solve a problem is no more use to the human species than a program which cannot solve that problem at all. What is missing is a way of measuring how long the solution of a problem is bound to take. This issue of computational complexity is at the centre of much recent mathematical research for very practical reasons. As soon as one adopts the computational scientists' point of view that the possibility of solving a problem within a reasonable period of time is the real issue, the theory of Turing machines loses much of its interest.

Suppose that we have a computer program written in one of the popular computer languages, and that it is of the form

```
line 1
line 2
...
line k
```

We use the word 'line' instead of the more technically correct 'executable state-ment' here and below. Then one can produce a new program from the old one by writing something similar to the following pseudocode.

```
timelimit = 10^10;
clock=0;
while clock<timelimit do
line 1
if clock<timelimit then clock=clock+1 else break;
line 2
if clock<timelimit then clock=clock+1 else break;
...
if clock<timelimit then clock=clock+1 else break;
```

```
line k
break;
end;
```

This program carries out exactly the same computation as the previous one with one extra feature. It counts how many steps it has implemented and if this reaches the number 10^{10} then it stops. Of course 10^{10} could be any other number, and in real computers one would probably set it so that the program would halt automatically within a few hours or days. Let us call a program with this particular structure a Program. Then anything which can be solved by a program can also be solved by a Program provided `timelimit` is large enough. However Turing's halting problem does not exist for Programs! If we have a Program then we know that it will stop within the time `timelimit`, and at that point we can see if it has provided the solution or proof which was sought. From this point of view programs are really infinite collections of Programs for which `timelimit` is allowed to have larger and larger values. The paradoxes relating to UTMs all arise because of this infinite character.

Of course the disappearance of the halting problem has not solved anything. Our Programs are less powerful than idealized programs which can run for arbitrary lengths of time. Nevertheless the discussion of Programs shifts one's attention from a semi-philosophical problem, namely is a program going to run for ever without coming to a conclusion, to one of more importance to humans, namely is the problem soluble within a useful time scale. It is interesting and perhaps surprising that automatic time limits are not a normal feature of programs. It is usually easier to write programs without such controls and simply to stop them if they have not finished within a reasonable length of time. The time limit is there, but it is implemented manually rather than automatically.

We turn next to computers which do not have time limits written into their programs, but which are restricted by having finite memories. Of course this is not actually a restriction! Let us define a FCM (finite computing machine) to be a machine with one or more finite sized processors (the chips which do the computations) and a finite, possibly very large, memory connected to the processors in any manner. The memory consists of a large number of sites, each of which can be ON or OFF, usually labelled 0 and 1. The state of the machine at any particular moment is the set of all values stored in the memory. It is clear that the number of possible states is finite, but even for current machines it turns out to be a huge number as defined in Chapter 3.

The processors are assumed to change the state of the machine at regular intervals of time, perhaps every nanosecond. They do this in a fixed manner, so that if the state at instant t is known then the state at instant $t + 1$ is completely determined by that. The problem is entered into the machine by setting up its initial state, so the initial state IS the problem (program). It then keeps on changing its state until it reaches a point at which its output device prints the solution and stops.

No matter how its processors operate, a FCM cannot keep moving to new states indefinitely since it only has a finite number of these. The maximum length of time to solve a problem cannot be greater than the total number of different states available multiplied by the clock time. If the machine is still going after that time it must be passing through some state for a second time, and must therefore be repeating itself rather than heading towards a solution. So we have a decision procedure, that is a means of knowing whether that machine is capable of solving that problem: wait for the relevant length of time and if the problem is not solved it never will be. On the other hand, even for current computers, the decision time is vastly longer than the lifetime of the Universe! So Turing's Halting Problem has again disappeared in the form he wrote it. The practical halting problem is still there: it may not be possible to tell whether a finite machine will solve a problem within a time scale of relevance to the human species, except by running it.

Passage to the Infinite

The nature of the infinite has caused more problems to mathematicians and philosophers than any other. Even Plato and Aristotle disagreed about whether the set of all integers exists actually or only potentially. The appearance of Cantor's set theory in the late nineteenth century seemed to resolve this problem, but developments in the first forty years of the twentieth century were to prove that many of the difficulties might never be resolved. Because most of these discussions revolved around logic and arithmetic, they had a discrete flavour which directed people's attention away from other attitudes towards the infinite.

Any finite set containing N points can be identified with the set of numbers n such that $1 \leq n \leq N$. These sets converge (in a sense which we need not make precise) to the set of all positive numbers as N tends to infinity. We have already seen that the difference between a finite computing machine and a Turing machine lies solely in the fact that the first has a large finite memory, while the second has an infinite memory, the memory sites being labelled by the positive numbers. The purpose of this section is to describe other ways of taking the infinite limit. These may be more appropriate in particular situations than simply passing to the set of all positive numbers. We will take statistical mechanics as an example of the appearance of new structures as one moves from finite to infinite systems.

The science of thermodynamics arose in the mid-nineteenth century as a result of the drive to improve the efficiency of steam engines. It describes the relationship between the bulk properties of gases, for example their density, temperature, and pressure. When we speak of phase transitions of bulk materials, we are thinking about their sudden changes of state as the temperature changes: water boils at $100\,^\circ$C (at normal pressures) and freezes at $0\,^\circ$C. Figure 5.8 shows how the state of a substance depends upon the temperature and pressure. Note that for high enough temperatures and pressures there is not a clear distinction

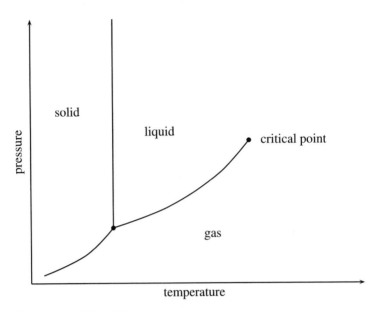

Fig. 5.8 A Generic Phase Diagram

to be made between liquids and gases. The critical point marks the point in the phase diagram at which this distinction disappears.

Thermodynamics is a phenomenological science, in that it does not make any reference to the fact that a substance is ultimately composed of individual molecules or atoms. This is achieved by statistical mechanics, whose goal is to explain the laws of thermodynamics starting from the interactions between individual atoms.

In statistical mechanics one starts with a finite collection of atoms, either quantum or classical, distributed randomly throughout a given region, with a known density. It is not obvious how to prove anything about the bulk properties of such an assembly, and the standard procedure is to take the limit of the system as the volume increases to infinity, keeping the density of particles constant. The infinite volume model is, paradoxically, easier to analyse than the more realistic finite volume case, but even with this simplification progress has been extremely slow. Nevertheless there are several special examples for which the existence of a phase transition has been proved with full rigour. In these cases the thermodynamic formulae can be derived from the atomic model by considering only global quantities, such as temperature and pressure. So the justification of thermodynamics involves two very hard steps: first the passage from a finite collection of particles to the infinite volume limit, and then the identification and analysis of the thermodynamics variables. The mathematics involved makes no use of *formal* set theory or *formal* logic. Of course it uses logic in the sense that mathematicians and physicists argue more or less logically, but nothing related to Turing or Gödel makes an appearance.

There is another way of seeing that different infinite limits of large finite sets may have entirely different structures. If points are laid out on a straight line at unit intervals starting at zero, then the infinite limit is clearly the set of all natural numbers. However, if they are more and more densely packed inside a unit square, then the appropriate infinite limit is the whole of the square, which has a continuum of points and a totally different geometry from the set of natural numbers. The same applies to computing machines. A parallel processing computer might be constructed by putting a computer chip, each with its own local memory, at each point of a two-dimensional square lattice, connecting each one to its nearest neighbours. If the lattice spacing is very small and the processors are correspondingly fast, then its behaviour might be modelled by a set of equations involving every point of the continuous unit square. These equations would of course be an idealized model of the real thing, but they might well be more useful than trying to describe it in Turing machine terms.

One might use this idea in the analysis of the functioning of the retina. This contains a large collection of neurons, but for image analysis it might be more useful to model it by a continuous plane region. Of course the retina is not a continuous system, but nor is it a Turing machine: it has chemical and biological parts, and does not have an infinite memory tape. The question is which of various mathematical models generates more insight into some aspect of its workings. The same issue arises when modelling the operation of the brain as a whole. As Philip Anderson wrote in a review of a book of Penrose:

> there is a fair amount of evidence that the mind is not a single, simple entity: it may be a number of independent autonomous units squabbling for a central dias. Multiple personality disorder is only an extreme form of what goes on in the mind all the time. There is no single Turing machine or single tape. It is not clear that it is correct to model a parallel collection of semi-independent machines that is in some sense wider than it is deep, in terms of a sequentially operating single algorithm. In discussing complexity, this can be a different 'large-N limit' with different capabilities.[19]

Are Humans Logical?

Arguments about the limitations of computers and UTMs are sometimes complicated by an almost mystical belief that in principle there are no limits on what human beings can understand. In Chapter 1 I presented the psychological evidence that our reasoning powers are both granted and constrained by the particular organization of our brains. One can profitably think of them as pattern recognition systems. Presented with a new scene, object, or even idea, our brains try to find the closest match out of all of the stored patterns. If an scene or idea is radically new total misunderstanding is very likely, because the closest stored pattern bears no relationship to what is being presented. The response of the brain to stimuli often has little to do with rational calculation, and can involve behaviour which people cannot prevent, even if they know it is inappropriate. Thus a spider sealed in a bottle or even seen on the television may be put in the 'dangerous spider' category and evoke a strong fear reaction, in spite of the fact that there is visibly no danger. There is substantial experimental evidence that our rational abilities are added on top of a quite different type of system, and function quite differently from the expert systems of computers.

We have seen that there are mathematical problems which are too hard for anyone to solve, but Goldvarg, Johnson-Laird, Byrne and collaborators have conducted series of experiments demonstrating the failure of people to solve even very simple puzzles correctly. I quote one typical example which involved a group of twenty Princeton students. The instructions emphasized the importance of the opening statement and the need to think carefully.

> *Only one of the following assertions is true about a hand of cards.*
> *There is a king in the hand, or an ace, or both.*
> *There is a queen in the hand, or an ace, or both.*
> *There is a jack in the hand, or a ten, or both.*
> *Is it possible that there is an ace in the hand?*
>
> *Nearly every participant in our experiment responded: 'yes'. But, it is an illusion. If there was an ace in the hand, then two of the premises would be true, contrary to the opening remark that only one of them is true.*

It appears that without thorough training in logic, people regularly fail to deal with such problems. The particular weakness which Goldvarg and Johnson-Laird identify is the inability to make correct mental models of false possibilities.[20]

Next a more complicated example. Albert and Bertrand are each given a card with a positive number written on it. They can look at their own card but not at the other one. They are told that the numbers differ by one, but they do not know which of them has the larger number. A is given the opportunity to declare what B has, or to pass (stay silent) if he is not sure. Then B is given a similar opportunity. The game continues until one of them declares the value of the other's card. Guesses are not permitted.

Let us start with a simple case. Suppose that B has the card numbered 2. Then he knows that A has either 1 or 3. If A has 1 then he will immediately declare that B has 2. So if A passes on his first turn then B can conclude that A does not have 1; B can thus immediately declare that A has 3. It is fairly easy to list all the cases in which the game terminates in the first round. They are

> *A has 1. A declares that B has 2.*
> *A has 2 and B has 1. A passes and B declares.*
> *A has 3 and B has 2. A passes and B declares.*

The simplest interesting case is when A has 4 and B has 3. I will give two different arguments leading to different conclusions, without saying which is right!

(i) Before they start playing both A and B can work out from their own cards that nothing can happen in the first round. That is both A and B must pass on their first turns. When this happens neither of them has learnt anything. Therefore the situation at the start of the second round is exactly what it was at the start of the first round. No progress has been made and the game cannot terminate.

(ii) The game starts with A passing, B passing and then A passing again. It is now the turn of B, who reasons as follows. A must have 2 or 4. If A had 2 then he would have started the game thinking that B had 1 or 3. In the former case A would have expected B to declare on B's first turn. He would have seen that B did not and concluded that B has 3. Therefore he would have declared on his (A's) second turn. But A did not do this. Therefore B can declare on B's second turn that A has 4.

One can go on and on analysing more and more complicated cases, with the possible conclusion that if A has a fairly large number then the two players pass many times before one of them declares. During this long period of passing, each of the players knows that nothing can happen, because each knows the value on the other's card to within an error of 2.

The paradox arises from feelings about whether such puzzles have real validity, and whether the other person will actually follow the rules. It may also depend upon our inability to construct mental models of other people's mental models of our mental models of their mental models of . . . , beyond a certain depth. A mathematician is trained to follow the logic wherever it leads, while the man in the street adopts a very different approach to problem solving. One could argue that in an extremely diverse world this attitude has a higher Darwinian survival value than the tightly structured response of the mathematician. If one of the main problems in social interactions is detecting deceit, then there it may be a bad idea to accept someone else's statements at face value, and more important to form an assessment of their character. The fact that inadequate methods of reasoning give the wrong answer to certain types of question does not matter if those questions rarely arise in the real world.

There is an analogy with the views of judges in a legal case and that of juries. The former clearly know more about the law and about legal arguments. Juries occasionally bring in 'perverse' verdicts which are contrary to the facts of the case and the rules of the legal system, because they value justice more than the rules of a particular legal system. This is regarded by some (but not by many judges) as one of the reasons for the retention of the jury system. However in highly complex financial fraud cases there are strong arguments for trusting the decisions to judges, because ordinary people are often out of their depth in such technical situations.

So it is with mathematicians; there are many areas in which we can and should trust their judgement, but that does not mean that we have to accept logical deduction as the only way of making decisions. In most situations which we deal with in ordinary life, we need to make repeated and rapid decisions subject to a constant flow of ill-defined new information about the external world. This is what our brains are evolved to cope with. To quote Anderson again:

> *(Penrose's) computers do not 'halt' until they have found an exact answer. This can be crippling. In the real world it is usually adequate to 'satisfice' to use Herb Simon's term. Methods directed at finding an acceptable way to do something can be much more efficient than exact ones. This is one way in which the mind can take advantage of its knowledge of the structure of the world.*[21]

A major difference between human discourse and computer languages is that in the former terms are learned by association rather than by being defined. For this reason, among others, natural language is not a good medium for conducting careful logical reasoning. There are rules of grammar within each natural language, but there are not such clear rules of interpretative correctness. Among the many well-known paradoxes I mention

> *The next sentence is true.*
> *The last sentence is false.*

As emphasized by Hofstadter,[22] each of the sentences is potentially useful on its own and only in combination are they deadly. Just as Russell introduced a theory of types to remove self-reference paradoxes in set theory, so one might introduce an infinite hierarchy of meta-languages to eliminate similar paradoxes in ordinary language. One can indeed prevent self-reference by insisting that one can only refer to a sentence in a meta-language which is at a higher level than the language or meta-language in which the sentence is written. Nevertheless, to resort to such a rule would be extremely limiting. It eliminates perfectly acceptable sentences such as

> *This sentence was typed in 2001.*
> *This sentence contains five words.*
> *This sentence is self-referential.*

It appears that either the theory of hierarchies of metalanguages is not the right way of eliminating the paradoxes of ordinary language, or natural language is incapable of being rendered consistent without being rebuilt from the ground upwards. One should not interpret the above as a criticism of natural language as an inferior mode of discourse. Indeed a literary figure might say exactly the opposite. The point is that the two are very different, and one should be very careful about being carried away by intricate arguments in natural language.

5.5   The Real Number System

Since the start of the seventeenth century, the concept of a real number has appeared to be unambiguous, even if a precise definition was not obvious. In this section we will see that the same difficulties arise as for integers, but in a worse form. I will argue that real numbers were devised by us to help us to construct models of the external world. Once again we start with a brief history of real numbers and some examples of the ways in which they are used. This will provide the evidence for the concluding discussion.

A Brief History

The discovery that the notion of length in geometry could not be developed using whole numbers and fractions (the only numbers they knew about) was made by the Pythagoreans in the fifth century BC. It caused a crisis in their closed group, and they tried to suppress the bad news. Little is known about this group of Greek mystics/geometers directly, and we have to rely upon sources such as Pappus and Proclus, writing in the fourth and fifth centuries AD respectively. It is known that they had access to much earlier documents, now lost. By the time of Euclid many proofs of key theorems in geometry had been discovered, but a proper theory of real numbers, as we now call them, still lay far in the future.

With the rapid development of navigation and astronomy in the sixteenth century, the need for efficient computational tools became steadily more urgent. At last the Indo-Arabic notation for manipulating fractional parts of numbers in the decimal notation had to be adopted. In 1585 Stevin published *De Thiende*, in which he described in detail the procedures for multiplying and carrying out other arithmetic operations with decimals. This notation was put to essential use by Napier quite soon afterwards (by the standards of those days). He was concerned to provide rapid and efficient methods of multiplying numbers and extracting square roots, and started work at the end of the sixteenth century on the production of tables for this purpose. His logarithm tables were published in 1614, near the end of his life, and their value was immediately appreciated by those who had previously spent long times on repetitive computations. The idea was taken further by Briggs, who published much improved tables of 10 digit logarithms in 1624. Such tables were in constant use right up to 1970, as were slide rules, which implemented the taking of logarithms mechanically using two sliding pieces of wood.

The use of decimals is readily associated with the idea that numbers may be identified with points on a continuous line, and this was to prove vital for Newton's development of the calculus later in the seventeenth century. The fundamental nature of this new type of 'real' number was, however, quite problematical, and it was not fully understood until the nineteenth century. A typical example of the kind of problems which arose concerns the geometric series

$$\frac{1}{1-x} = 1 + x + x^2 + x^3 + x^4 + \cdots.$$

One can check the particular case

$$2 = 1 + \tfrac{1}{2} + \tfrac{1}{4} + \tfrac{1}{8} + \tfrac{1}{16} + \cdots$$

by adding together the first dozen terms on the right-hand side. However, Euler and others in the eighteenth century were willing to put $x = 2$ to deduce the nonsensical

$$-1 = 1 + 2 + 4 + 8 + 16 + \cdots.$$

Mathematicians might have argued that they should be allowed to use this formula in the middle of a calculation, since similar manipulations with the equally absurd $i = \sqrt{-1}$ always led to valid results. One problem for us in understanding such a formula is that we interpret it in the light of an understanding of real numbers which only emerged much later.

One of the principal people to put analysis on a firm foundation was Cauchy. In the 1820s he gave a precise definition of convergence and proposed a programme which would determine when a formula which involved adding together an infinite number of terms was acceptable. However, this still left the precise nature of real numbers unclarified. Several different but equivalent definitions of the concept of real number resolved this problem around 1872. While this resolved the foundational problems of analysis, it did so at the cost of making the real numbers into axiomatically defined objects, whose relationship with the intuitive idea of points on a continuous line required a considerable effort to understand. Indeed Dedekind referred to his definition as *creating* the real numbers. From this time onwards many mathematicians concluded that analysis was a matter of formal proofs, and that their geometrical intuition was a guide to the truth, but not an infallible one. This aspect of the definitions of real numbers was deeply regretted by some. For example, du Bois-Raymond (1882) wrote about it 'demeaning Analysis to a mere game with symbols', while Heine, one of the inventors of the new approach, wrote:

> *I take in my definition a purely formal point of view, calling some symbols numbers, so that the existence of these numbers is beyond doubt.*

There is an aspect of the above story which must not be forgotten. If the geometrical idea that real numbers can be thought of as points on a continuous line had not existed, nobody would have tried to carry out the rigorous construction. Concentrating on the formal side of the final product is like admiring one of Shakespeare's plays on the basis of his extraordinarily large vocabulary. In both cases there is a deeper way of judging what has been produced, which goes beyond merely technical issues. In mathematics formal constructions are only of interest if they correspond to some degree with earlier intuitive ideas. This involves a judgement which can only be made outside the formal system by a human being. The construction of the real numbers confirmed what was already believed, but also went further in allowing mathematicians to resolve questions which had not previously been imagined. While the above construction of real numbers is accepted as the best working tool in most circumstances, there are other formalizations, such as non-standard analysis and constructive analysis, which are of value in certain contexts. None of them can claim to be 'the correct' way of formalizing our intuitive notions of the continuous line, any more than one can say words are 'the correct' way of communicating ideas. This may be true in the great majority of situations but it is not a necessary truth in all.

The result of the above developments was to give mathematicians confidence that their previous more intuitive ideas did not harbour some hidden inconsistencies. They continued to rely upon their geometrical image of the real line as they always had, and used the technical definition simply to reinforce or supplement the geometrical picture when this was needed.

The peculiarities of the new analysis were soon to be demonstrated. In 1872 Weierstrass showed that it was possible to draw curves which did not just have changes of direction at a few points, but which had no direction (tangent line) at any point on them: by peering more and more closely at the curves one could see that they had infinitely many corners. Figure 5.9 shows one such curve, which arises in hyperbolic geometry and is described in more detail in the notes.[23]

Such curves were considered by Poincaré, Hermite, and many of the other prominent mathematicians of the late nineteenth century to be pathologies of the worst kind—lacking computer graphics they could not see how beautiful they were. Following the efforts of Mandelbrot, we have eventually come to terms with these possibilities, and have even constructed a new science of fractal

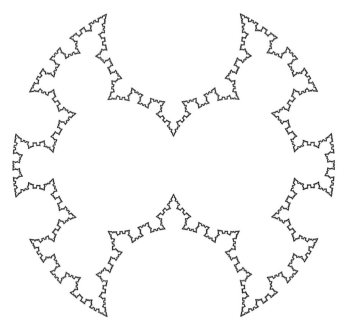

Fig. 5.9 A Curve without Tangents
By permission of David Wright, using software developed for 'Indra's Pearls'

objects around them. In other words we have developed a new geometrical intuition, in which these objects appear natural rather than pathological. But it should not be forgotten how many people were profoundly dismayed at the consequences of definitions which they had chosen to accept.

Classical mathematics contains much worse peculiarities than directionless curves. The Tarski–Banach paradox of 1924 has several forms, but the easiest to understand is the theorem that a sphere can be broken up into a finite numbers of parts which may then be moved around and reassembled to create two new spheres, each the size of the original one! The weasel words are 'can be'. The theorem does not refer to the real world in which spheres are made of atoms and the total number of atoms is conserved. It refers to a mathematical model of reality including an axiom asserting the existence of certain exotic sets which nobody could possibly construct. So much the worse for this axiom you might think, but the mathematical community currently thinks otherwise. Fortunately it is a free world.

What is Equality?

A standard issue in logic, much discussed by philosophers, is the law of the excluded middle, (LEM). This is the claim that every meaningful statement is either true or false, and that the only issue is to find out which of these is the case. If one accepts this and certain aspects of set theory, then it is a consequence of the work of Gödel and Turing that there are statements which are true but not provable by any Turing machine.

In the case of a definite statement about the numbers (integers), the LEM seems to be unarguably correct if one believes that the numbers exist in some independent sense, on the grounds that one must concede that a meaningful statement about an independently existing entity has a truth value. If, however, the numbers only exist by a social convention then it is also a matter of convention whether one chooses to use the law of the excluded middle. The same issues apply to real numbers, but even more so. The standard constructions of the real numbers from the integers assume that there is no fundamental issue involved in asserting that two real numbers are or are not equal. Whether there is a means of resolving this question in a particular case is taken to be a practical problem. The goal of this section is to demonstrate that issues connected with the equality of two real numbers lead to some interesting paradoxes.

Let us define a real number a as follows. We start by putting $a = 5/9$ and write it on top of π, both in decimal notation.

$$a = 0.55555555555555555555\ldots$$

$$\pi = 3.14159265358979323846\ldots$$

We now alter a and π systematically as follows. Looking through π, if we find any sequence of a thousand or more consecutive digits which are all the same, we switch those digits with the corresponding digits of a. This produces two new numbers, which we call b and σ to avoid confusion with the previous ones. The number b is as well-defined as almost any number in mathematics. We can compute each of its digits in a finite length of time by a well understood and routine procedure, and so can calculate b as accurately as we like. Let us now think about the exact value of b. If no sequence of a thousand consecutive identical digits occurs in the decimal expansion of π then nothing happens and $b = \frac{5}{9}$. However if a sequence of a thousand consecutive identical digits does occur then whether $b < \frac{5}{9}$ or $b > \frac{5}{9}$ depends upon which particular digit is repeated a thousand times first.

It seems extremely difficult to imagine how the occurrence of such a sequence (of a thousand successive identical digits) might be proved or disproved, so we may never know whether $b = \frac{5}{9}$ or not. But let us suppose that one day someone proves a general theorem about the randomness of the digits of π with the implication that such a sequence does occur somewhere in its decimal expansion. It is even more unlikely that the *first* such sequence will ever be discovered. So we could then be in the highly embarrassing situation in which we knew that b was not equal to $b = \frac{5}{9}$ but did not know whether it was greater than or less than $b = \frac{5}{9}$. Certainly its difference from $b = \frac{5}{9}$ would be so small as to be invisible in any practical sense.

In Platonic mathematics such numbers as b exist and therefore either are or are not equal to $b = \frac{5}{9}$. This belief, however, is without value in settling which of these possibilities occurs. The Platonic philosophy may be comforting, but it does not carry the subject any further forwards. One can plausibly argue that science is concerned with finding and presenting evidence and not with discussing philosophical views about the nature of truth. Intuitionists have developed this idea in a systematic manner, as we shall now see.

Constructive Analysis

The school of intuitionist mathematics was dominated by Brouwer during the 1920s, but it was also strongly supported by Hermann Weyl. Brouwer advocated a constructive approach to the subject, in which mathematical entities would only be regarded as existing once an effective construction for them had been written down. Such ideas had already been expressed forcibly late in the nineteenth century by Kronecker, who had advocated the rejection of most of analysis and set theory.

The intuitionist programme entailed the removal of the law of the excluded middle (LEM) from the mathematician's toolbox. According to Brouwer:

> The long belief in the universal validity of the principle of the excluded third in mathematics is considered by intuitionism as a phenomenon of the history of civilization of the same kind as the old-time belief in the rationality of π or in the rotation of the firmament on an axis passing through the earth. And intuitionism tries to explain the long persistence of this dogma by two facts:

firstly the obvious non-contradictority of the principle for an arbitrary single assertion; secondly the practical validity of the whole of classical logic for an extensive group of simple everyday phenomena.[24]

The above statement flows from Brouwer's anti-Platonism and in particular his belief, following Aristotle, that the set of numbers does not exist as a completed entity in itself. One should think rather of a process for producing numbers which can be continued indefinitely. Bernays explained it as follows:

This point of departure carries with it the other divergences, in particular those concerning the application and interpretation of logical forms: neither a general judgement about integers nor a judgement of existence can be interpreted as expressing a property of the series of numbers. A general theorem about numbers is to be regarded as a sort of prediction that a property will present itself for each construction of a number; and the affirmation of the existence of a number with a certain property is interpreted as an incomplete communication of a more precise proposition indicating a particular number having the property in question or a method of obtaining such a number.[25]

Brouwer's philosophy of mathematics was rejected by most other mathematicians, partly because of his difficult personality, but mainly because it entailed the loss of some of the most important branches of mathematics. There are even intuitionist theorems which are definitely false in classical mathematics.[26]

Brouwer's intuitionism (INT) is only one of several constructive approaches to mathematics. In 1967 Errett Bishop developed an approach which avoided the main problems of INT. The simplest summary is that he avoided the use of the law of the excluded middle, but did not replace it by an alternative. A consequence is that every theorem of Bishop's constructive mathematics (BISH) is also a theorem of classical mathematics. In the reverse direction some theorems of classical mathematics either do not appear in BISH or appear in a modified form. Contrary to the doubts of the sceptics, Bishop proved that one could develop a large part of analysis within such a context by actually writing out the details of the proofs. In order to prove that an equation has a solution within BISH one needs to give a method for evaluating that solution. It is not sufficient to derive a contradiction from the assumption that no solution exists.

If one wanted to explain BISH in one sentence it would be as follows. In classical mathematics ∃ means 'there exists' but in BISH it means 'the writer has an explicit way of producing'. Thus in BISH one accepts that mathematical statements are made within a particular social context and what cannot be asserted today might very well be asserted in the future. One focuses not on whether statements might be true in some Platonic sense but on whether we have a method of proving them. Everything else is a result of following up this idea systematically.

One does not need to make any philosophical commitment in order to take an interest in BISH. In particular one does not have to believe that the law of the excluded middle is false or meaningless. Avoiding its use can be regarded merely as a way of forcing one to concentrate on the constructive aspects of classical mathematics. Inevitably constructive proofs are harder than traditional proofs: they provide more information, and one never gets something for nothing. Although Bishop's version of mathematics takes some getting used to, it does have real interest for any mathematician who has even a small respect for computational issues.

Non-standard Analysis

At the end of the eighteenth century mathematicians were using two ideas which caused them great unease. We have already discussed the first, the status of imaginary numbers. The second has an equally interesting history. It is the nature of infinitesimals, introduced by Leibniz in the late seventeenth century in his development of the calculus. These were supposed to be infinitely (or perhaps indefinitely) small but non-zero numbers denoted dx, such that either $dx \times dx$ vanishes or it can be deleted from calculations without 'essential error'. Of course the meaning of the phrase 'essential error' caused mathematicians some anxiety, but the results obtained using Leibniz's notation were so valuable that they felt unable to abandon its use. The status of these infinitesimals was apparently resolved by their abolition in the 1820s, when Cauchy gave a new and rigorous definition of limit. This provided proper foundations for the calculus without ever mentioning infinitesimals.[27] For the next hundred years infinitesimals were universally agreed to have been one of the necessary mistakes in the development of mathematics.

Unfortunately for those who like their history simple, and who would like to think of mathematics as the gradual unveiling of some objective reality, the situation was to reverse yet again. In 1961 Abraham Robinson pioneered yet another approach, called non-standard analysis, in which the real number system is augmented, rather than diminished as in constructive analysis. In this system there do indeed exist a variety of infinitely big and infinitely small but non-zero 'numbers', and Robinson developed a systematic way of doing analysis with these infinitesimals. After four hundred years Leibniz's notation for differentiation at last makes sense! The system is consistent with the standard system in the sense that any theorem about 'traditional' real numbers proved using non-standard analysis can also be proved by classical methods. The classical proof may, however, appear very unnatural. Non-standard analysis has a slowly increasing number of devotees, and has recently been used to provide a more intuitive proof of the Jordan curve theorem, a result about the geometry of the plane. It has also been responsible for new mainstream theorems, particularly in the area of probability theory called stochastic processes. Its philosophical and historical importance has been acknowledged by a variety of mathematicians including Lakatos, Bishop, and Kochen. It is one of the clear cases of a revolution in mathematical thought, albeit one which is being played out over a period of about fifty years.

5.6 The Computer Revolution

Identifying a revolution while it is still in its early stages is a fool's game, but I will take the risk. I believe that the rapid growth of computer power is already leading to changes in the way mathematicians work, and that within fifty years the impact will be enormous.

To develop this idea, let me go back to 1900. If one looks Cambridge Tripos examination papers, one sees that students were expected to have a mastery of special functions—Bessel functions and their relationships for example. Even in 1930, when my father was in university, the study of such topics occupied a substantial fraction of a typical degree course. When I went to university in 1962, I had already learned by rote dozens of formulae involving trigonometric functions, and had spent many hours carrying out calculations involving them. However, Bessel functions had more or less disappeared from the compulsory part of the Oxford University syllabus. By that time there were several weighty volumes listing their properties and providing tables to compute them. People who needed to use functions named after Bessel, Struve, Airy, Whittaker, Riemann, and many other mathematicians referred to the volumes as needed. The quantity of information is these books was far beyond human memory, and anyway we had more interesting things to think about.

Today's students know only the basic addition formulae for trigonometric functions when they arrive in university, if indeed that much. Many university departments now start their courses teaching students how to use Maple or Mathematica, software written by specialists and containing vast arrays of formulae. If one needs to differentiate, evaluate or find the zeros of a Bessel function, one now only needs to type in the correct command to get the answer. The next generation will use this software, not knowing where the formulae come from, and assuming that the computer is always right.

This is certainly a loss, but the gains are considerably greater. For example the trigonometric function $\tan(x)$ can be written as a power series in x. It takes only a few seconds to ask Maple to write out the first ten terms of the expansion

$$
\begin{aligned}
\tan(x) = {} & x + 1/3 * x^3 + 2/15 * x^5 + 17/315 * x^7 + 62/2835 * x^9 \\
& + 1382/155925 * x^{11} + 21844/6081075 * x^{13} \\
& + 929569/638512875 * x^{15} + 6404582/10854718875 * x^{17} \\
& + 443861162/1856156927625 * x^{19} + O(x^{20})
\end{aligned}
$$

a task which would have taken me hours or, more likely, days.

It seems inevitable that as time passes more and more of our knowledge will be integrated into a universal online system. Every theorem will have its own internet address with links to all of the results on which it depends. New ideas will be tested automatically for consistency with already accepted facts, with apparent conflicts referred to humans for resolution. Formulae involving special functions will be confirmed by evaluating them against thousands of randomly chosen numbers. Theorems will be assigned reliability weightings by computers which monitor the number of other results which tend to confirm them, and the number of mathematicians who have used them without objection. These weightings will factor in the authority of the mathematicians who discovered or used the theorems.

A few years ago there was a vigorous debate in the Bulletin of the American Mathematical Society about whether it was necessary to have a method of quality assurance at all.[28] There were many who believed that the most important breakthroughs in the subject were made by people who were working entirely intuitively, leaving the details of theorem-proving to others. In fact, of course, one needs both generals and soldiers. Generals can be inspired, but they can also live in a fantasy world which is exposed when others try implement their grand plans. The automation of some of the work of the soldiers is inevitable and does not threaten the generals; indeed it may enable them to access the resources which they need more efficiently.

Is this prospect frightening? Well, did the transfer from walking to horses and then to motor cars involve a diminution of our humanity? It is the same question. It may worry *us*, but our children will take it for granted, because they know nothing else.

Discussion

The origins of the real number system are deeply enmeshed with the belief that the world is in some deep sense continuous. This has a biological basis associated with our physical size, and is certainly scientifically incorrect once one gets down to the atomic level. During the second half of the twentieth century the definitions of our units of measurement have gradually acknowledged the discreteness of matter. Since 1967 the second has been defined as $9, 192, 631, 770$ times the period of a specific transition of caesium-133, and since 1983 the metre has been defined as the distance light travels (in a vacuum) in one second, divided by $299, 792, 458$. The kilogram is still the mass of a particular platinum-iridium cylinder kept near Paris. It is easy to imagine it being re-defined as the mass of a certain number of hydrogen atoms, but changing the definition will depend upon basic advances in technology. There is a current prospect of providing a new fundamental standard of electrical current based upon quantum theory, and hence on counting.

One of the strongest arguments for the independent existence of real numbers is that they are indispensable for understanding the physical world. Very unfairly, I have altered Shapiro's presentation of the argument in one important respect, discussed below.

1. Real analysis refers to, and has variables that range over, abstract objects called 'real numbers'. Moreover one who accepts the truth of the axioms of real analysis is committed to the existence of these abstract entities.
2. Real analysis is indispensable for fluid mechanics. That is modern fluid mechanics can neither be formulated nor practised without statements of real analysis.

3. If real analysis is indispensable for fluid mechanics, then one who accepts fluid mechanics as true is thereby committed to the truth of real analysis.

4. Fluid mechanics is true, or nearly true.

The conclusion of the argument is that real numbers exist.[29]

Unfortunately the final assumption is questionable. We do not actually believe that fluid mechanics is true. It is highly *accurate* in many circumstances, but fluids are actually composed of atoms, and these are discrete. The accuracy of fluid mechanics is eventually a result of a different theory, statistical mechanics, which has a completely different mathematical structure. If we replace truth by accuracy, then the best which the above argument can yield is that real numbers are a very useful tool. They may exist, but the argument does not prove this.

Now I come to my change in Shapiro's argument. Where I have written fluid mechanics Shapiro wrote physics. It seems bold to suggest that the whole of physics is not true, and that its ability to make accurate predictions of a huge range of phenomena is not evidence for this. This, nevertheless, is what I do. There have been major revolutions in what we regard as the fundamental theory, and currently we know that we do not have one. Many physicists are willing to contemplate the idea that space-time is actually discrete, or even that it is totally unlike our present ideas about it. It has to be admitted that all current theories are formulated in continuous terms. Current researchers use a variety of sophisticated tools from differential equations to quantum mechanics and Riemannian geometry, but the mathematics involved is just as continuous as was Euclidean geometry. Whether the world itself is ultimately continuous or not is unknown, and cannot be decided on the basis of the properties of our current models of it. The models change with time, and we have no idea what they may be like a hundred years hence. Maybe all of our current mathematical models will be replaced by a theory of cellular automata, as Stephen Wolfram has recently proposed. Only time will tell.

Notes and References

[1] See Propositions 21 to 28 of Archimedes' *On the Sphere and Cylinder, I.*

[2] A set is a collection of entities, called its elements, which may have been specified by means of some common property.

[3] Ewald, 1996, p. 1159

[4] Ewald, 1996, p. 1165

[5] Dummett 1978, p. 214

[6] Cohen 1971

[7] Ruelle 1998

[8] Lakatos 1978, p. 23

[9] This metaphor seems to be due to Quine.

[10] See page 266 for a discussion of empiricism in science.

[11] Gillies 2000a

[12] Davies 2002

[13] Turing 1950

[14] Gödel actually said little about his beliefs, but an unpublished document of his is discussed in detail by Wilfrid Hodges. See Hodges 1996.

[15] Church 1936–7

[16] Earman and Norton 1996, Hogarth 1996

[17] Davies 2001

[18] Blum et al. 1989, Traub and Werschulz 1998

[19] Anderson 1994

[20] Goldvarg and Johnson-Laird 2000

[21] Anderson 1994

[22] Hofstadter 1980, p. 21

[23] The figure is the limit set of a 'quasi-fuchsian' discrete group generated by two fractional linear transformations a and b and their inverses. The generators are determined up to conjugacy by the conditions $\text{trace}[a] = 2.1$, $\text{trace}[b] = 2.1$, $\text{trace}[aba^{-1}b^{-1}] = 0$. The figure is homeomorphic to the circle but has no tangent points. We thank Dave Wright for providing and explaining the figure, and refer to Mumford et al. 2002 for many other beautiful pictures and a much fuller account of the mathematics involved.

[24] Benacerraf and Putnam 1964, p. 82

[25] Bernays 1964, p. 278

[26] For example according to Brouwer *every* function $f : \mathbf{R} \rightarrow \mathbf{R}$ is continuous.

[27] The credit for the new definition of limit cannot be given solely to Cauchy, but his book *Cours d'Analyse* was a major point in the systematization of the new ideas.

[28] Stöltzner 2001

[29] Shapiro 2000, p. 228

6
Mechanics and Astronomy

6.1 Seventeenth Century Astronomy

This chapter considers two related topics. The first is the development of astronomy in the sixteenth and seventeenth century, culminating in Newton's publication of his laws of motion in 1687; the second concerns the subsequent history of these laws. More and more observations confirmed the predictions of Newton's theory, and after about 1750 nobody had any doubt that his theory of gravitation provided a *true* description of the world. The task of philosophers was to explain how finite human beings were able to acquire such certain knowledge of the world. Then in the first decades of the twentieth century it was discovered that this certainty was a chimera. Einstein dethroned Newton, and physics moved into a period of flux which has continued ever since.

The fact that such a well-established theory could eventually be superseded poses a severe challenge to any theory of scientific knowledge. We re-tell the story of the period, selecting the aspects which are most relevant to this matter. In the second half of the chapter we then describe some of the developments which led to the collapse of the Newtonian world-view. Finally in Chapter 10, we will resolve the problem by invoking the modern distinction between reality and mathematical models of reality.

The seventeenth century marks a decisive break between a social system dominated by the authority of the Roman Church and the rise of a more individualistic study of the world. At the start of the century the Church dominated Europe and claimed the right to interpret scientific findings.[1] By the end it was known that the motions of the planets were controlled by Newton's laws, that is by impersonal mathematical equations. The influence of these laws has never waned. In the twentieth century space craft have navigated around the solar system using Newton's laws to guide them, and the collision of the comet Shoemaker–Levy 9 with Jupiter in July 1994 was predicted months in advance by the same laws.

Nevertheless, the first quarter of the twentieth century was to bring two fundamental scientific revolutions, each of which totally changed scientists' view of the nature of the physical world. At the atomic level Newton's laws were to be replaced by quantum theory and at the cosmological level Einstein's general theory of relativity was to supersede them. These developments will be described later, but in this chapter we will concentrate on internal difficulties arising from the laws themselves. Although some hints of serious problems appeared late in the nineteenth century, it was only in the second half of the twentieth century that scientists realized their extent. The occasion was a dramatic increase in our ability to carry out extremely lengthy calculations by using computers. It led to the discovery of chaos—the highly unstable dependence of the solutions of Newton's equations on the initial conditions. This might be regarded as no more than a tiresome computational limitation. However, it is now recognized that it affects most of the phenomena in the real world of interest to us. Following Popper, I will argue that to believe that Newton's laws are still applicable 'in principle' in chaotic situations is to make a philosophical choice. This may be defended by reference to Ockham's razor, but it cannot be supported by scientific evidence. Indeed to predict the movement of real particles in chaotic situations, one would need to include effects due to quantum mechanics. But Newton's laws demonstrate their own limits quite independently of the appearance of quantum theory and general relativity.

The story starts during the second century AD, when the Alexandrian astronomer Ptolemy elaborated the earlier ideas of Hipparchus into his famous Ptolomaic system, described in *He mathematike syntaxis*. In this system the Earth was at the centre of the universe, and the planets moved around it in complicated orbits described in terms of cycles and epicycles. The Ptolomaic system survived for over a thousand years, and blended conveniently with the official dogma of the Church. Eventually Nicolaus Copernicus' book *De Revolutionibus*, proposed a model of the solar system in which the Sun was at the centre and the planets rotated around it. His system still involved the use of cycles and epicycles, but it was nevertheless substantially simpler and more convincing. Among its revolutionary proposals was the idea that the stars were not embedded in a crystal sphere which rotated around the Earth, but that they were stationary and the Earth itself rotated around an axis through its poles. Copernicus was well aware that his ideas might provoke outrage, and postponed its publication until 1543, the year of his death. He dedicated it to Pope Paul III in a letter which submitted entirely to the superior judgement of the Pope.

De Revolutionibus was not published by Copernicus himself, but by a friend and Lutheran theologian called Osiander, who was justifiably concerned by the fact that Luther had condemned such ideas as contrary to writings in the Bible. He added an anonymous introductory letter explaining that the book did not claim to be a true theory, but merely a method of calculation, to the extreme anger of those who had entrusted him with its printing. This letter might have contributed to the Roman Church tolerating its existence throughout the sixteenth century.

It would be nice to report that Copernicus' ideas were immediately accepted, at least by the astronomical community. Unfortunately history is rarely so simple. His work was read, but did not attract enormous attention because it was regarded as physically implausible, in spite of its conceptual simplicity. Later in the sixteenth century Tycho Brahe put forward a third model of the universe, in which the Sun and Moon moved around the Earth, while all of the planets moved around the Sun. We would now say that they were describing the same theory, but that Brahe preferred to use a rotating Earth-centred coordinate system rather than the simpler Sun-centred coordinates. We would regard this as a perverse (but not incorrect) choice, because it makes the equations more complicated. However, Brahe had good reasons for thinking that it was the physically correct choice.

Although the two theories were the same as far as the motions of the planets were concerned, Brahe considered that if the Earth was truly moving around the Sun, this would have visible effects on the apparent positions of the stars. Namely, as the Earth moved around its orbit, they would appear to move slightly—an effect called stellar parallax. This effect has now been observed, but it is extremely small, and far beyond the discrimination of Brahe's instruments. It is an interesting comment on the history of science that Copernicus' theory was accepted long before this problem had been resolved experimentally, simply by supposing that the stars were far more remote than had previously been thought. The difficulty was brushed underneath the carpet, but history has confirmed that this was actually the right thing to do.

An understanding of the religious context of the times is of vital importance in explaining the Church's later reaction to Galileo. In 1520 a long simmering conflict between the Church and Luther came to a head. He was threatened with excommunication unless he withdrew his increasingly strident criticisms of Papal indulgences and other degenerate activities of the Church leadership in Rome. This spurred him on to open rebellion and the threatened excommunication took place in September of that year. Luther responded by casting the Papal bull into the fire at Wittenberg in December and declaring that nobody could be saved unless he renounced the rule of the Papacy. Luther spent the rest of the decade fomenting a successful nationalist Protestant rebellion against Papal rule in Germany. During the next decade Calvin led a Reformation movement in Geneva. Henry VIII, who was more interested in power than in doctrine, took control over the English Church in Britain, and was duly excommunicated in 1538, when he started the process of dissolving the monasteries and acquiring their very substantial assets for his own use.

All of these events understandably caused a crisis in the Roman Church, which desperately needed to carry out internal reforms and re-assert its authority. This it did in the Council of Trent, which in 1546 laid out rules limiting the rights of any individual to object to its teaching on grounds of conscience. It also gave the Church the final authority in all matters concerning the interpretation of the Bible. This can be illustrated by the following circular of the Jesuits:

> *Let no one defend or teach anything opposed, detracting or unfavourable to the faith, either in philosophy or in theology. Let no one defend anything against the axioms received by the philosophers, such as: there are only four kinds of causes; there are only four elements; there are only three principles of natural things; fire is hot and dry; air is humid and moist . . . Let all professors conform to these prescriptions; let them say nothing against the propositions*

here announced, either in public or in private; under no pretext, not even that of piety or truth, should they teach anything other than that these texts are established and defined. This is not just an admonition, but a teaching that we impose.

Thereafter the Church was extremely sensitive to any further suggestions of revolt, and was prepared to impose its own decisions ruthlessly when it considered this necessary. In 1600 Giordano Bruno was burned at the stake in Rome for having advocated over a period of years the physical correctness of the heliocentric theory of Copernicus and the even more radical idea that the our Sun and planets might be just one of a large number of similar systems spread throughout an infinite universe.

Bruno's problem was not simply to be ahead of his time. He was incautious to the point of absurdity. At the same time as promoting his cosmological views throughout Europe, he advocated a religious doctrine (theosophism) which bore little resemblance to Christianity, let alone to Catholicism. When tried for heresy, he made no attempt to recant, even after seven years in prison, and refused to withdraw or even moderate any of his claims. The conclusion was inevitable.

Galileo

Galileo Galilei spent much of his life investigating the principles governing the behaviour of bodies such as balances, levers, pendulums, and falling weights. This was a very confused subject at the time: notions such as force and inertia were not well understood, and many mathematical tools which we now take for granted did not exist until after he had died. Indeed Galileo regarded geometry as the language of mathematics, rather than algebraic equations. The story about his dropping bodies of various weights from the leaning Tower of Pisa is famous. Unfortunately it appears nowhere in his own writings, and is probably a later invention of his biographer Viviani.

The Tower of Pisa story, nevertheless, encapsulates in a vivid image something true and important: that Galileo overturned the scholastic myth proclaiming that heavier bodies fell faster in proportion to their weights. In fact, as Galileo explained at length in *Dialogue Concerning the Two Chief World Systems*, bodies of different weight fall at the same speed, once one has discounted the effects of air resistance.

Galileo's approach to science has been absorbed into our culture so thoroughly that it is hard for us to appreciate its revolutionary nature. He insisted on the importance of experimental or observational evidence, and that one should not accept the word of authority, whether this meant the scholastic philosophers or the very powerful Church in Rome. His explanations of phenomena were not always correct, but his discoveries provided the crucial context for Newton's later work. Of course he was not alone: there was a rising class of mechanically skilled workers who became increasing self-confident as the sixteenth century progressed, and he inherited their view about the right way to acquire knowledge.

Galileo is most famous for his astronomical discoveries and subsequent conflict with the Roman Church, but he has another equally important claim to fame. Mechanical clocks had existed since medieval times, the earliest surviving example which still works being that at Salisbury Cathedral, dating from about 1386. Galileo's idea of regulating such clocks using a pendulum was to transform their accuracy. In 1659 Viviani produced a drawing of a simple pendulum clock which had been designed by Galileo shortly before his death in 1642. Huygens, however, actually built such a clock in 1657. From this point onwards progress was extremely rapid. By 1725 it had led to the introduction of temperature compensation, watches with spring balance regulators, and a variety of other ingenious ideas. Clocks were then accurate to better than one second per day, and further innovations were to continue until atomic clocks arrived in the twentieth century. An enormous amount has been written about the social origins of the explosion of science in the seventeenth century, but if one had to pick out the most important single contribution, it might well be the invention of the pendulum clock.

We now turn to Galileo's astronomical discoveries. The story starts when Roger Bacon wrote about spectacles in the mid-thirteenth century; convex lens spectacles were already being manufactured in Florence by 1300.[2] The telescope may well have been invented by Hans Lippershey, a spectacle maker from the Netherlands. Its military and commercial value were first recognized in 1608. Galileo learned of it very soon after that and started making his own improved copies in Florence. This was not an easy task: he had to grind his own lenses and work out how to improve the optics in order to get higher magnifications.

Galileo made his main astronomical discoveries in 1610. He examined the surface of the Moon in detail, observing irregularities on it which he correctly interpreted as mountains; he was even able to estimate that the height of one mountain was at least six kilometres. These observations flew in the face of the entire body of scholastic understanding of the heavenly bodies, which were believed to be perfect. In January he first saw four moons of Jupiter. His initial interpretation of them was as points oscillating back and forth along a straight line, but he soon re-interpreted what he could see as rotation in circular orbits seen edge on. He soon started to write his book *The Starry Messenger*, which included drawings of craters and mountains on the Moon. Published in March 1610, it was an instant best-seller, and prompted Kepler to write his own pamphlet *Conversations with the Starry Messenger*, emphasizing the importance of Galileo's observations.

There were, however, others who were not willing to accept Galileo's ideas immediately. In 1610 most telescopes were of very poor quality, and this resulted in suggestions by some scholastic philosophers that the more controversial objects which he claimed to have seen were illusions produced by his instruments.[3] Galileo actually *saw* dark patches on the surface of the Moon whose size and shape changed according to the phase of the Moon. He realized that the changes were what one would expect if sunlight was striking the tops of mountains and causing shadows, and made this *interpretation*.

Fig. 6.1 Far Side of the Moon
Reproduced from http://spaceflight.nasa.gov/images/

It is very easy to dismiss the objections of the scholastic philosophers as wholly misguided, but Galileo could be completely wrong, even on important matters. (The same is true of Newton, Darwin, Hilbert, and Einstein.) He rejected the view of Brahe and others that comets were material bodies, arguing instead that they were merely optical phenomena. He was right about the Moon, not because of his superior logic, but because his interpretation of the evidence provided by his telescope was later confirmed by a wide range of independent evidence. This includes many wonderful NASA photographs, such as figure 6.1. Hindsight is a wonderful thing, and is frequently associated with a simplification of old disputes, in which the losers are presented as rather stupid, while the winners are endowed with god-like powers of insight.

Galileo did not come out in favour of the Copernican theory in *The Starry Messenger*, but stronger evidence was not long in coming. In the autumn of 1610 Venus was in the right position for him to be able to confirm Copernicus' prediction that Venus should exhibit phases, a phenomenon which was entirely inexplicable within the scholastic tradition. The phases of Venus were (and still are) explained by assuming that Venus shone by reflected light and that it orbited the Sun at a distance less than that of the Earth from the Sun. When it was approximately between the Earth and Sun, it was at its biggest and also appeared only as a narrow crescent. On the other hand when on the opposite side of the Sun from the Earth, it was at its smallest and appeared as a fully illuminated disc.

When Galileo turned his telescope to the Sun he saw that there were dark spots or patches on its surface, which gradually changed shape as well as moving from one side of the Sun to the other before disappearing at the edge. Galileo concluded that the Sun was imperfect, and that it rotated about an axis. In other words the Sun was also a material object, extraordinary only because it shines by its own light. He had numerous exchanges with others about the nature of the sunspots, and wrote that they bore some similarity to clouds, but was careful not to go beyond what he could see.

Galileo was not the first to see the spots on the Sun, even within Europe. Indeed, their existence had been known to the Chinese for about two thousand years, but they did not invest this fact with any deep religious or philosophical importance. The main beneficiaries of the Chinese passion for recording astronomical events were twentieth century astronomers, who found their detailed records of eclipses and other unusual events enormously valuable. For Galileo, the existence of sunspots was to be one more piece in the argument leading to the overthrow of the scholastic philosophy of the heavens.

The news about Galileo's astonishing discoveries spread very rapidly and copies of his telescope were sent all over Europe. As a result he became the most celebrated philosopher in Europe, and also a public advocate of the Copernican theory. This started a conflict between Galileo and the Church (more precisely certain powerful people within it). The opposition to Galileo came partly from the fact that the Copernican theory was in conflict with what was in the Bible. However, Augustine had emphasized many centuries earlier how important it was not to interpret Biblical statements about the natural world too dogmatically, lest the Church might later come to appear foolish. Galileo was scathingly rude about some eminent supporters of the Aristotelian system, and, unsurprisingly, some of them were keen to take revenge. An example of the sharpness of his attacks is provided by the following passage from the later *Dialogue Concerning the Two Chief World Systems*:[4]

> *The anatomist showed that the great trunk of nerves, leaving the brain and passing through the nape, extended on down the spine and then branched out through the whole body, and that only a single strand as fine as a thread arrived at the heart. Turning to a gentleman whom he knew to be a Peripatetic philosopher, and on whose account he had been exhibiting and demonstrating everything with unusual care, he asked this man whether he was at last satisfied and convinced that the nerves originated in the brain and not in the heart. The philosopher, after considering for awhile, answered: 'You have made me see this matter so plainly and palpably that if Aristotle's text were not contrary to it, stating clearly that the nerves originate in the heart, I should be forced to admit it to be true.'*[5]

Other influential figures in the Church hierarchy considered that Galileo was trying to reduce the Church's authority by arguing too dogmatically for his views. He was trying to take control of the terms of the debate, and claiming the right to interpret theologians. He, on the other hand, desperately wanted to disseminate his ideas concerning the physical correctness of the Copernican theory. Unlike Bruno, he tried very hard to present his ideas as being in conformity with Church teaching, particularly that of Augustine, but ultimately to no avail. In 1616 Copernicus' book was banned, and Galileo was instructed not to promote his ideas about the truth of the Copernican theory.

Galileo accepted from Aristotelian philosophers the distinction between theories which had only been proved mathematically (i.e. were computational aids but not literally true) and those which had been proved physically (i.e. demonstrated beyond reasonable doubt). This distinction was clearly relevant to the Ptolomaic system, which provided an accurate but completely artificial mathematical scheme for predicting the motion of the planets across the sky. It was argued by the Aristotelians that the Copernican scheme was merely a simpler and better system of the same type. Foucault's pendulum experiment proved the rotation of the Earth beyond reasonable doubt, and could have been performed by Galileo, but was not in fact carried out until the mid-nineteenth century. The plane in which such a pendulum swings rotates slowly; the rate depends upon the latitude of the site, and is easily explained as a consequence of the rotation of the Earth. If the Earth was stationary, it is difficult to imagine any explanation of this effect.

Galileo tried hard to find evidence for the rotation of the Earth which did not rely upon gazing into the heavens, and believed that he had found this in the existence of tides. Unfortunately, his theory of the tides was flawed. He claimed that they were a consequence of the combined effects of the Earth's rotation and motion around the Sun, an idea which we now see to have been based upon his imperfect understanding of dynamics. Many others, from Antigonus in ancient Greece to Yü Ching in China, had correctly suggested that they depended primarily upon the influence of the Moon, but Galileo dismissed this idea with contempt in *Dialogue Concerning the Two Chief World Systems*. It is much easier for us to see his arrogance when he dismissed the influence of the Moon as not being worthy of serious discussion, than it is on matters about which history has shown him to be right.

In fact direct evidence for the rotation of the Earth *can* be obtained by studying the oceans. The rotation produces an effective force, called the Coriolis force, which has a profound effect on the circulation patterns of both oceans and atmosphere at a global level. It also provides the reason why hurricanes in the northern hemisphere always rotate anticlockwise. Unfortunately systematic information of this type was not available to Galileo, nor would he have been able to make sense of it before Newton's laws had been discovered.

Galileo did not stop his investigations into the Copernican theory after 1616. Matters came to a head after the publication of his book *Dialogue Concerning the Two Chief World Systems* in 1632. He adopted the device of presenting his ideas in the form of a debate between three characters, so that he could claim that their views were not his own. But it was obvious where his own sympathies lay, and in 1633 he was finally charged with heresy and forced, under threat of torture, to recant publicly. The Church seemed finally to have won, but the Copernican system was so much simpler, and the evidence so easily available to anyone with a telescope, that by 1650 his observations were widely regarded as overwhelming evidence for its physical correctness. The eventual acceptance of Newton's theory of gravitation finally made the Church's official theory an historical irrelevance, but it was not until 1992 that Pope John Paul II officially apologized for the Catholic Church's error in persecuting Galileo. He admitted that Galileo had been right both scientifically and theologically, but could not bring himself to declare that the Church itself had been at fault, preferring to blame theologians whose minds had been insufficiently flexible. This may be technically correct, but there is no doubt that Pope Urban VIII had fully supported the conviction and subsequent house arrest of Galileo.

One of the important passages in the *Dialogue* was Galileo's lengthy discussion of objections by Brahe and others to Copernicus' theory. Would not the rotation of the Earth prevent objects such as leaves falling through the air vertically, as the Earth moved underneath them? Would not it have the (demonstrably wrong) implication that a cannon ball fired to the west would carry further than one fired to the east? Galileo answered this in a famous passage about a person in a cabin of a ship which is moving with constant speed. He pointed out that fish in a bowl were able to swim around with complete freedom, and that water drops emptying from a bottle fell vertically when measured by reference to the cabin. In other words all motions in the cabin were the same whether or not the cabin was moving, *provided* they were measured with respect to the cabin. Galileo concluded that observations of falling objects could not be used as an argument against the rotation of the Earth.

This is a completely non-mathematical argument, but is nevertheless a *principle of relativity* based on an argument of the same type as was later used by Einstein. It was later transformed by Isaac Newton into the first of his three laws of motion. It should not, however, be imagined that Galileo fully understood all the implications of his own ideas. He remained wedded to the idea that bodies naturally moved in circular paths. The circular orbit of the Moon around the Earth therefore did not need an explanation in terms of gravitational forces: it was only doing what came naturally to it. Rather than being smug about our superior insight, let us consider how our descendants in four hundred years time will think about our own failures to see the obvious!

Kepler

Johannes Kepler was born in 1571 to a poor family, but was fortunate to win a scholarship to study in the Lutheran seminary at the University of Tübingen. He was taught astronomy by Maestlin, who persuaded him privately that the Copernican system was true, even though he was teaching the Ptolomaic system publicly. As I have explained, its physical truth was not accepted by the Church, and it was not a safe doctrine to advocate in the political and religious turmoil of those times. As a result of his early promise, Kepler was invited to join the staff of the astronomer Tycho Brahe in Prague, and soon succeeded him as Imperial Mathematician to the Holy Roman Emperor Rudolf II in 1601, when Tycho died. He then had access to the vast quantity of detailed observations by Tycho, and spent years trying to find a unifying mathematical explanation of them. In 1609 this culminated with the publication of his first two laws of planetary motion in *Astronomia Nova*. Ten years later he published his third law in *Harmonice Mundi*. They stated that:

> *The planets move around the Sun in elliptical orbits with the Sun at one focus.*
>
> *A line from the Sun to a planet sweeps out equal areas in equal times.*
>
> *The squares of the orbital periods of the planets are proportional to the cubes of their mean distance from the Sun.*

The first law was revolutionary: it abolished the special status of circles in the description of planetary motion, and replaced them by ellipses. The second described how rapidly a planet moves at different parts of its orbit. The final law explained how to relate the orbital periods of the different planets to each other.

Kepler's book *Epitome of Copernican Astronomy*, published in 1618–21, was provocatively titled, and this was rewarded by its being put on the Index by the Church. Kepler benefited from the relatively great intellectual freedom of Rudolf's court in early years, but later had to contend with a series of political and religious pressures. This culminated in his prolonged defence of his mother against a charge of witchcraft between 1615–21.

It is interesting that Galileo did not pursue Kepler's ideas concerning elliptic orbits, even though they corresponded on several occasions. Probably he regarded ellipses as being as mystical and unscientific as Kepler's introduction first of the Platonic solids, and later of musical harmonies, into the description of the Solar System. In addition Galileo was heavily involved in promulgating the Copernican theory, supported by telescopic observations rather than mathematical formulations. When Newton published his *Principia* in 1687, he also avoided references to Kepler, in spite of the fact that he derived Kepler's laws from his own theory of gravitation.[6]

There is another reason why Kepler's laws might have been ignored by many of his contemporaries. Kepler derived them by a long search for the simplest mathematical formulae which would reproduce the planetary orbits, but the result did not relate to anything else known about the planets. They appeared to *describe* reality, but did not *explain* anything, and could only be verified by someone who was willing to spend months re-analyzing the data. Of course the preferred description in terms of circles did not explain anything either, but at least circles were familiar and simple. Perhaps this illustrates the fact that science is (or should be) about explanation rather than prediction.

Newton

The culmination of seventeenth century research in astronomy and mechanics came with the work of Isaac Newton. Born in 1642 a few months after the death of his father, in relatively modest circumstances, a series of lucky accidents enabled him to go to Trinity College, Cambridge as a student in 1661. As with so many other great scientists, his most fundamental work was done at an early age, but he did not publish his laws of motion for many years. One of the reasons was that he was misled by inaccurate astronomical data into doubting its complete success. Eventually he was persuaded by Halley to publish *Philosophiae Naturalis Principia Mathematica* in three volumes in 1687, to acclaim among the very few equipped to understand it. *Principia* was written in Latin, as was almost all scientific work until the nineteenth century. He wrote it in a severe classical Euclidean style, in order, so he wrote, 'to avoid being bated by little smatterers in mathematics'. Among the consequences was a widespread lack of awareness of the magnitude of his achievement, which persisted for many years.

Principia starts with the three laws of motion, which are rendered rather freely below:

> *Every body continues in its state of rest, or of uniform motion in a straight line, unless acted on by an external force.*
>
> *The acceleration of a body is proportional to the applied force and is in the same direction as the force.*
>
> *The actions of two bodies on each other are equal but opposite in direction.*

In a Scholium immediately following, Newton stated his indebtedness for these laws to Galileo (but not to Kepler or Descartes), and mentioned later developments by Wren, Wallace, and Huygens. In fact Newton went far beyond Galileo in the clarity of his understanding of dynamics, and it is probable that he was not familiar with Galileo's actual writings. He should instead have acknowledged his debt to Descartes, but would not have done so because of his disagreements with the Cartesians. *Principia* is not important just for the formulation of the above laws. Its key feature was the development of his theory of gravitation. His deduction of the inverse square law of gravitation from 'the phenomena' is rightly considered to be one of the scientific triumphs of all time.

Although scientists in the seventeenth century did not regard detailed references to earlier work as de rigeur, Newton's attitude in this respect was quite extreme. He was extremely jealous of his reputation, and liked to give the impression that he owed little of substance to anyone else. He deliberately removed references to Hooke from *Principia*, because of the latter's earlier (but unsupported) claim to have proved the inverse square law for gravitational forces. Even his famous statement 'If I have seen further it is by standing on ye shoulders of Giants' was likely meant as a subtle insult to the stooped and physically deformed Hooke. Later in his life he used his position as President of the Royal Society of London[7] ruthlessly to try to establish that Leibniz had stolen his work in calculus. This was in fact entirely untrue.

The Law of Universal Gravitation

Like other early members of the Royal Society, Newton claimed to follow Francis Bacon's ideas about the proper way to do science. Rule 4 in Book Three appears to summarize this method very simply:

> *In experimental philosophy, propositions gathered from phenomena by induction should be considered either exactly or very nearly true notwithstanding any contrary hypotheses, until yet other phenomena make such propositions either more exact or liable to exceptions.*[8]

When one examines the arguments used in *Principia*, one finds a much more complicated picture. In a famous General Scholium added to *Principia* in 1713 Newton emphasized that he was led from observations to laws by a process of induction, and rejected the use of hypotheses which might result 'from dreams and vague fictions of our own devising'. This was not, as some people have claimed, a rejection of the use of properly formulated hypotheses which could then be tested, but an attack on the Cartesian school. When Newton came to studying the orbits of comets, he explicitly assumed that they were parabolic, used observations to determine the precise orbits under this hypothesis, and then confirmed the validity of the conclusions by means of further observations. In fact Newton was an opportunist. He used any and every method, as was convenient.

Newton did not just show that the inverse square law of gravitation explained all of the observed phenomena. He gave two independent arguments using the astronomical data which *proved* that the forces on bodies in the solar system *must* obey the inverse square law. He did not use the elliptical character of orbits in either proof, presumably because planetary orbits are so close to circular that distinguishing between ellipses and other curves is not a straightforward matter. The first of his proofs depended on Kepler's third law, but his second proof was much more decisive.

When a planet moves in its orbit around the Sun, there are certain easily measured features of its orbit which depend sensitively upon the force law. These refer to the apsides, the points on the orbit at which the planet is closest to or furthest from the Sun. Newton proved that for approximately circular orbits the positions of these apsides changed from one orbit to the next *except* in the case of the inverse square law. In fact he treated a number of different possible gravitational laws in considerable detail. Figure 6.2 shows the angle A between two apsides for an orbit which is *not* controlled by the inverse square law; the circles of closest approach and furthest distance from the centre of gravitation are shown as dotted lines.

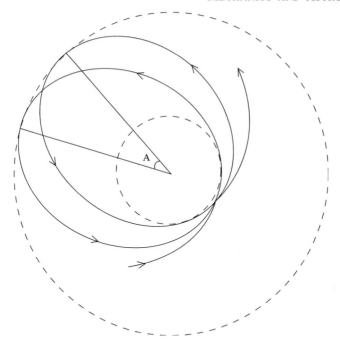

Fig. 6.2 Orbit with Rotating Apsides

Newton was then able to deduce Proposition 2 of Book 3, namely:

The forces by which the primary planets are continually drawn away from rectilinear motions and are maintained in their respective orbits are directed to the sun and are inversely as the squares of their distances from its centre.

... But this second part of the proposition is proved with the greatest exactness from the fact that the aphelia are at rest. For the slightest departure from the ratio of the square would (by book 1, prop. 45, corol. 1) necessarily result in a noticeable motion of the apsides in a single revolution and an immense such motion in many revolutions.

The orbit of the Moon did not quite fit this law, and Newton discussed several possible reasons in *Principia*. However, even in this case an acceptance of the data without corrections would have forced him to change the power only slightly, from 2 to 2 · 0165. He correctly judged that the accumulation of all of the evidence justified the conclusion that the inverse square law applied to *all* bodies.

The decision to examine the 'quiescence of the apsides' was inspired. Newton used a combination of ideas from his newly invented calculus, but developed in as geometrical a manner as possible. On the observational side, however, he only needed the 'null' observation that the apsides did not move from one orbit to another. This could be confirmed with great accuracy because one could observe a planet over many orbits. One might say that never has so much been deduced from so little!

Newton's law of universal gravitation included the assertion that the gravitational force between two bodies depended upon their masses, but not upon the type of substance they were made of. This argument involved two steps, the first of which was a *thought experiment* designed to convince the readers of *Principia* that the force between planets was the same as the familiar weight which we experience on the surface of the Earth. Today we have overwhelming proof of this via satellites and space probes, but Newton had no comparable evidence. He argued that if the two forces were distinct, then a satellite orbiting the Earth just above the highest mountains would be subject to both, and that the consequences would be implausible. He completed the argument with a long series of pendulum experiments, concluding that the composition of the weight did not make any difference to the period of the pendulum, if one compensated for the effects of air resistance. This fact was not entirely novel, but Newton's experimental design was very clever, and enabled him to prove the result to far higher accuracy than had previously been possible, in spite of having no precise method of measuring time.

The idea that gravity might depend upon the substances involved was re-investigated fifteen years ago, when there appeared to be evidence of a short-range correction to Newton's law of gravitation, called a 'fifth force'. After several repetitions by others of the experiment of Peter Thieberger in 1987, the present consensus is that no such force exists.

In spite of its success, the Newtonian theory had one very unsettling feature. Newton's laws described two distant bodies attracting each other by a gravitational force whose strength and direction depends on where they were with respect to each other. No explanation was given for how either of the bodies could be aware of the existence of the other, when there was nothing between them but empty space, let alone how they could know how far away the other body was and in what direction. To put it less anthropomorphically, Newton proposed no mechanism by which this remotely generated force could arise.

For Huygens and several others this was an unacceptable weakness of his theory. Many seventeenth century scientists accepted Descartes' argument that everything in mechanics should be explained in terms of material interactions between bodies which were in contact. His 'explanation' of the orbits of the planets involved their being carried around the sun in a swirling vortex of some ethereal fluid. Newton put a considerable effort into proving that this idea could not work. It might be plausible for planets whose orbits were almost circular, but comets were known to follow quite different types of orbit, which intersected the planetary orbits at substantial angles. It was simply not possible to construct a coherent account of how any fluid could produce the variety of orbits observed.

Newton addressed this issue in his General Scholium of 1713. He stated that he adopted no hypotheses concerning the reasons for the gravitational force, but contented himself with the fact that gravity did really exist. In retrospect we may regard this as a defining moment in the history of science. It marked the time after which scientists started to admit steadily more non-material entities, gravitational and later electric and magnetic fields, whose existence could only be inferred from their effects. It also led to the idea that a scientific 'explanation' might be nothing more than the formulation of mathematical equations which yielded correct predictions. This incursion of advanced mathematics into physics was destined to continue without pause up to the present day. A consequence is that much of physics is now incomprehensible to all except a very few.

Newton is now considered to have been one of the greatest geniuses of all time, ranking with Archimedes and Einstein, but his theory of gravitation was not fully accepted until several decades after his death. Our judgement of him depends upon forgetting his profound interest in alchemy, and the energy he put into studying the Old Testament and its chronology. He actually spent far more of his life on these subjects than he did on *Principia*. As with so many geniuses, his outstanding characteristic was an ability to commit himself totally to a single issue for as long as it took. Sometimes the results were worthwhile, and sometimes they were not.

6.2   Laplace and Determinism

From our current point of view one of the key facts about this scientific revolution was that it abolished our special status in the universe. Coming to terms with this was a slow process—indeed it is still going on. Before the seventeenth century the Christian world was unashamedly centred on a theological view of man as the centre of God's act of creation. Afterwards scientific laws based upon cold mathematics became ever more powerful. It is not surprising that people came to believe that there were no limits to the power of the new science, and that it provided a complete description of reality. One must remember how small people's control over their world was before the seventeenth century, and how steady was the chain of new discoveries over the next three hundred years.

During the eighteenth century, Newton's theory was developed much further by several French mathematicians, using Leibniz's calculus. In spite of the efforts of Euler and d'Alembert, the anomaly in the orbit of the Moon remained, and by 1750 they were ready to declare that Newton's inverse square law would have to be modified. Then Alexis-Claude Clairaut proved, in 1752, that the difficult calculation of the gravitational interactions between the Moon, Earth, and Sun had been done incorrectly; he demonstrated that the inverse square law did indeed yield the observed motion of the Moon. He followed this up by refining an earlier prediction by Halley of the return of a comet, now named after him. Clairaut's calculation that it would reappear in April 1759 was only a month late—a success so dramatic that it could not be due to chance.

The person who made the greatest contribution to the detailed analysis of planetary orbits was Pierre-Simon, marquis de Laplace (1749–1827). He gave complete explanations of nearly all the outstanding anomalies in planetary orbits. The success of his programme persuaded any remaining sceptics that Newton's theory constituted the final word on this subject.

Laplace was also responsible for the clearest expression of the deterministic principle. He proposed that if a sufficiently vast intellect had complete knowledge of the positions and velocities of all the bodies in the Universe at some instant, then it would be possible for it to work out the exact subsequent motion of every body indefinitely into the future.[9] Since the positions and velocities do have values, even if we do not know them, this means that our future actions are pre-ordained. These ideas were very influential, and were not effectively challenged until the twentieth century, when their philosophical and scientific bases were both undermined fatally.

In *The Open Universe* Karl Popper argued that one should distinguish between metaphysical and scientific determinism. The former refers to how things actually are without reference to whether we could possibly have any evidence for them. If God exists and knows exactly what we will do at every moment in the future, then our free will is an illusion and the future is completely determined. Whether or not the future course of events is controlled by scientific laws or mathematical equations is a completely separate issue.

On the other hand, scientific determinism refers to whether a being *who is a part of the universe*, like ourselves but far more powerful, might be able to predict the future with arbitrarily specified accuracy provided sufficiently accurate data about the present were obtained. Popper argues convincingly that this is not possible for a variety of reasons. Among these is the impossibility of gathering data of sufficient accuracy to predict the future of a complex multi-body system, and the fact that in some situations effects associated with the person making the prediction cannot be disregarded in the way that Laplace assumed. This might not be relevant in astronomy, but astronomy concerns systems which are far simpler than most of those to which Newtonian mechanics might in principle be applied.

Since the time of Laplace (many) physicists have changed their attitude towards scientific laws. Rather than being true representations of reality, they are considered to be mathematical models, with limited domains of applicability. A model may fail to be useful either because the equations are not exact for some values of the relevant parameters, or because the equations cannot be solved (by us). In the first case we can legitimately look for a better mathematical model, but in the second it is possible that *no modifications* of the model will be amenable to computation. In the next few sections we show that such situations do indeed exist, and that there are ultimate limits to the mathematical method.

Chaos in the Solar System

Although the phenomenon of chaos is usually regarded as a recent discovery, it has quite a long history. The first occasion on which a chaotic physical system was described may well have been in a lecture by a Professor Pierce to the British

Association in 1861, recorded in their Yearbook. He wrote down the formulae governing the motion of a pendulum when the point of suspension is made to move uniformly in a vertical circle. 'He then exhibited beautifully executed diagrams on transparent cloth, which showed by curves, some most regular and some most fantastic in their forms, the behaviour of such a pendulum under various conditions'. He also demonstrated the high instability of the irregular motions in his lecture. We now know that this was a genuine instance of chaotic dynamics, but his observations were not followed up by others.

The next important development occurred in 1890, when the mathematician Henri Poincaré submitted a memoire to a prize competition held under the patronage of King Oscar II of Sweden and Norway. Its subject was the solution of what we now call the restricted three-body problem for bodies moving under Newton's laws of motion. Poincaré mentioned the example of a small planet whose orbital motion around the Sun is perturbed by Jupiter. Shortly before the publication of his memoire, Poincaré became aware of a serious error in his arguments, and had to withdraw the few copies which had already been circulated. The version which was eventually published was substantially different from the one which won the prize, although he tried to minimize the significance of this fact. He had discovered the existence of very peculiar orbits, which seemed to defy simple description. Eventually he wrote in *Les Méthodes Nouvelles de la Mécanique Céleste, vol. III*, published in 1899, that:

> One is struck by the complexity of this figure, which I am not even attempting to draw. Nothing can give us a better idea of the complexity of the three-body problem and in general all the problems of dynamics where there is no single-valued integral and Bohlin's series diverge.

In this book Poincaré had stumbled on the phenomenon of chaos. A precise definition of this is beyond the scope of this book, but the key idea is that of massive instability. If one throws two stones in a very similar manner then one expects them to follow very similar paths. In chaotic dynamics this is only true for a limited time: the deviations between the two paths builds up so rapidly that after a certain length of time they become effectively independent, *however close they originally were*. A simple analogy is with a pinball game, in which one knows that the trajectory of the ball depends entirely upon how fast it starts, but it is almost impossible to control where it goes in spite of this. Slightly more precisely if one compares the motion of a body starting from two very slightly different initial positions, the difference between the positions doubles over a short time interval. Thus after ten time intervals the difference has multiplied by a thousand, after twenty by a million and before too long the two trajectories are completely unrelated to each other. The important point is that it is impossible in practical terms to predict where the body will be even a short time into the future, however accurate the measurement of its initial position is.

Although the importance of Poincaré's discovery of chaos was soon apparent to Hadamard, Duhem, and others, it was largely neglected until the 1950s because of the impossibility of carrying out quantitative investigations. Computer technology has advanced so far since those days that simple computer programs which demonstrate chaos for the restricted three-body problem may now be found in elementary textbooks. Starting with mathematical work of Kolmogorov in 1954 and more applied work of Lorenz in 1963 on meteorology, this is now one of the most active areas of mathematical research.

One of the exciting discoveries in the astronomy of the Solar System in the last few decades has been that chaotic behaviour of the type which Poincaré discovered may actually be observed. It operates on a timescale of a hundred thousand years or more, but nevertheless has dramatic consequences. Between the Sun and the roughly circular orbit of Jupiter, there is a region containing thousands of asteroids with a great variety of sizes and orbital periods. When one counts the number of asteroids with each particular orbital period, and compares this with the orbital period of Jupiter, one makes a surprising discovery. There are no asteroids to be found whose orbital period is a third of that of Jupiter, an absence referred to as a Kirkwood gap.

Following extensive and difficult numerical computations, the reason for this absence has now been found. If one computes the orbital behaviour of an asteroid whose period is almost exactly one third that of Jupiter, it appears to be reasonably stable over periods of tens of thousands of years. On the other hand every hundred thousand years or more its orbit suddenly changes dramatically, taking it much closer to the Sun for a short period before returning to its previous form. To explain why this should imply that there are no asteroids with such orbits, we have to bring Mars into the picture. When its orbit takes it closer to the Sun an asteroid may pass inside the orbit of Mars. It has a small chance of colliding with or having its orbit dramatically changed by that planet. Over a long enough period of time all such asteroids will be removed from the belt. Although Mars comes into the picture at the end, the chaotic nature of the orbits of the asteroids is a consequence of solving Newton's equations for three bodies, the asteroid, Jupiter and the Sun, as Poincaré had foreseen.

Hyperion

In many situations in Solar astronomy one can do computations as if the planets and their satellites are point objects. There are two reasons for this. The first is that it can be shown that the gravitational field of a spherical body is the same as it would be for a point mass. The second is that the distances between Solar bodies are usually so much larger than their diameters that small deviations from sphericity are not significant.

When one examines the orbits of many satellites one discovers a phenomenon called spin-orbit coupling. Our own Moon moves once around the Earth every twenty eight days, which is exactly the same as the time it takes to rotate once about its axis. The result is that the same face of the Moon always points towards the Earth. The reason for this behaviour is that the Moon is not exactly spherical and the eventual result of the gravitational forces acting on it has been as just stated.

The phenomenon, called 1:1 spin-orbit coupling, is common to many of the satellites in the Solar System. Mercury is unique in having what is called 3:2 spin-orbit coupling with respect to the Sun. This means that Mercury takes exactly 2/3 of an orbit, about 59 Earth days, to rotate once about its axis. This is the length of the sidereal day on Mercury, defined by measuring rotation by reference to the distant stars.

The satellite Hyperion of Saturn is an unimpressive lump of rock about three hundred kilometres across. It was discovered in 1848, and photographed by Voyager 2 in 1981. Not only is it very irregular in shape, but the eccentricity of its orbit is also unusually large.[10] The result of these peculiarities is that instead of rotating stably around an axis, it tumbles as it orbits Saturn. The irregularities of this tumbling have been observed from the variations in its light curve as viewed from Earth, and during the fly-by of the Voyager 2 probe. Mathematical models have been constructed to explain what is happening. They show that one may predict the tumbling motion fairly accurately for times up to its orbital period of $21\frac{1}{4}$ days. Over longer periods the model becomes more and more unstable, and after a year it is impossible to make any meaningful predictions; further analysis shows that the mathematical model is chaotic over a very large region in its phase space. It is generally agreed that this is not just a feature of the model, but describes how Hyperion does actually tumble in its orbit.[11] So we are in the interesting situation in which the application of Newton's laws yields the conclusion that it is not possible to use them to make accurate predictions of the motion of Hyperion one year into the future.

Molecular Chaos

Let us consider a gas of air molecules at room temperature and pressure, occupying a box which is a one metre cube. In this situation there are about 10^{25} molecules in the box. Let us suppose that the molecules are reflected by the walls of the box without loss of energy, and that there are no influences on them from outside the box. The average distance travelled between collisions is about 200 times the diameter of a molecule. From the point of view of a molecule they are widely separated from each other and move long distances between collisions. From our point of view the situation is quite otherwise: each molecule is involved in over a billion collisions per second!

The molecules are considered to move freely in straight lines and to bounce elastically when they collide with each other. They are also considered to obey Newton's laws. Actually of course their collisions should be described by quantum mechanics, but this would not resolve the paradox to be discussed. Now suppose that we could solve Newton's laws for some typical initial Configuration 1 of the molecules, and let us compare the result with that of another Configuration 2. We suppose that in Configuration 2 every particle except one has exactly the same initial position and velocity, and that one is displaced by an extraordinarily small amount, say one trillionth of the diameter of a molecule. When this molecule has its first subsequent collision, it emerges with a slightly changed direction, perhaps one trillionth of a degree different. The slight change in direction implies that by the time it hits the next molecule its displacement is substantially bigger than it was before it hit the first molecule; let us suppose that at each collision the displacement is multiplied by a factor of two. After about 50 collisions the molecule is displaced by more than the diameter of a molecule, so the next collision does not take place. From this point, which certainly takes less than a millionth of a second, the evolution of the molecules in the two gases becomes rapidly more different. The effects would be apparent at a macroscopic level within a minute or so, if one could find a way of measuring them.

We have been discussing the effect of a tiny change in the initial conditions of the gas, but the same effect arises if one makes a tiny change in the evolution law, such as would be caused by influences of the outside world on the gas. Even if one knew the initial position and velocity of each particle in the box perfectly, the gravitational influence of very distant bodies will rapidly lead to a change in the detailed movement of the molecules of the gas. Indeed only about a hundred collisions are needed before the gravitational influence of an electron at the furthest limits of the Universe has a substantial effect on the motion of the molecules! In other words the notion of a gas in an isolated box is a fiction. No part of the universe can in practice be isolated from any of the rest!

The above is written as if it were possible to know everything about the initial state of the molecules, apart from one detail which leads to the rapid appearance of major changes everywhere in the gas. However, the real situation is far worse: one does not know the exact position or velocity of *any* of the molecules. The result is that it is hopeless to think that the motion of the molecules can be predicted in any but a statistical sense. This may be sufficient for many purposes, but it is not always so. On page 172 we will discuss observations made by the botanist Brown which depend entirely upon this random motion of the molecules.

The behaviour of fluids in bulk appears to depend little upon their constituent molecules. Fluids are studied as if they were continuous distributions of matter, using differential equation methods. Sometimes the solutions of these equations are well behaved, as when describing the smooth flow of water down a channel. Under other circumstances the fluid flow is turbulent, and the relevant differential equations are impossible to solve with any accuracy. It is quite possible that in such situations the continuum description of fluids is not a valid approximation: in other words the molecular nature of fluids may indeed be important in the turbulent regime.

This a recent insight. Forty years ago it was thought that as computers became more powerful and the mathematical modelling more precise, it would be possible to produce reliable weather forecasts further and further into the future. Unfortunately in 1963 the meteorologist Edward Lorenz discovered that *extremely simple* equations of this type can exhibit chaotic behaviour, just as Newton's equations can. His idea is often summed up in the now famous motto: the flap of a butterfly's wings in Brazil may set off a tornado in Texas. This way of describing the effect of chaos was not originally due to Lorenz, but he did adopt it. It is picturesque but quite wrong. There is no way of distinguishing the effects due to the butterfly from trillions of other small movements which might equally well be regarded as 'causing' the tornado. One cannot perform two calculations, in one of which the butterfly flaps its wings and in the other of which it does not: the very instability of atmospheric dynamics renders such a computation inconceivable. They could only be done by a being with an (almost) infinitely powerful computer who does not have any material interactions with the universe at all!

The phenomenon of chaos establishes that accurate long range weather forecasts are in principle impossible. It does not, however, mean that there is no point in trying to improve the models used for weather forecasting. When considering a period of a few days model error might well be more important than the error associated with chaos. A practical consequence of Lorenz's discovery is that meteorologists now make not one weather forecast, but a whole series based upon very slightly varying initial data. Sometimes the results are very similar to each other and one can have confidence in the forecast. On other occasions they can be quite different and one can only give probabilities for the various forecasts.

A Trip to Infinity

The following extra-ordinary scenario using Newton's equations was recently discovered by Xia.[12] It has nothing to do with chaos, but illustrates a different way in which Newton's laws can break down. Consider five bodies: two orbit closely around each other in one position while another two orbit around each other in a second position, as in figure 6.3. The fifth, which we call the particle, oscillates back and forth between the two pairs, passing exactly half way between the bodies of each pair at either end. When it is not between the pairs, they both pull it back; it eventually stops and moves back towards them.

There is a small but very important difference between successive oscillations of the particle. Each time it passes a pair of bodies its gravitational attraction pulls them inwards towards it, with the result that they lose potential energy and move more rapidly. Some of this increased speed is passed on to the particle.

Xia's clever idea is as follows. He showed that it is possible to choose the initial positions and masses so that the particle moves back and forth between the two pairs of bodies faster and faster while the two pairs accelerate away from each other. The process happens ever faster, and both pairs of bodies disappear to infinity *in a finite length of time*! The particle shuttles back and forth, blurring out into a continuous line which stretches infinitely far in both directions!

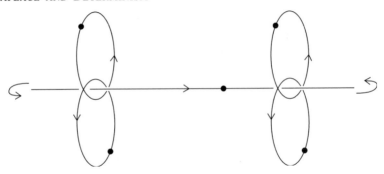

Fig. 6.3 Xia's Five Body Phenomenon

Of course the above would never happen in the real world. The whole point of the example is to emphasize the difference between a *mathematical model* and *physical reality*. The mathematical conclusion is irrelevant because at sufficiently large speeds Newton's theory must be replaced by general relativity. Secondly, in order to acquire the kinetic energy to achieve the effect, the orbiting pairs must move steadily closer to each other. Eventually one has to confront the fact that the bodies cannot be point particles and must collide. On the other hand the model uses Newton's laws in exactly the same way as that used to predict planetary orbits, except that we start with rather unusual initial conditions. In this problem Newton's equations *have no solution* after a finite period of time, even though there are no collisions between the bodies concerned.

The bizarre scenario described above raises questions about the Solar System. It appears to be stable, but we already know that the asteroids show chaotic behaviour. How can we know whether solving Newton's equations may not lead to the Earth being ejected from the Solar System some distant time in the future? The equations cannot be solved numerically if one looks far enough ahead (many millions of years) and we have no guarantee that Newton's equation will protect us for ever.

The Theory of Relativity

I have delayed mentioning the theory of relativity because I wanted to emphasize that Newton's laws of motion themselves show that they cannot provide a complete description of reality. There are situations in which the relevant computations are impossible because of chaos, and others in which there exist no solutions to the relevant equations. Historically the crisis for Newton's theory came not from the discovery of chaos but from Einstein's theories of special and general relativity. According to Einstein himself, the development of electromagnetic theory was of key importance in this story. This started when Faraday uncovered the close connection between electricity and magnetism by a brilliant series of experiments in the first half of the nineteenth century.

At that time it was natural for scientists to seek to explain electromagnetic phenomena in mechanical terms. They posited the existence of a substance called aether, which filled the whole of space, and hoped that electric and magnetic fields might be explained as elastic distortions and vibrations of the aether. However, it gradually became clear that the properties of the aether would have to be so peculiar than it would hardly qualify as a substance. This approach to the subject was abandoned after Maxwell discovered his electromagnetic field equations at King's College, London between 1860 and 1865. His prediction of the possibility of creating electromagnetic (i.e. radio) waves in the laboratory was verified by Hertz in 1886. When Henri Poincaré, one of the world's leading mathematical physicists, wrote *Science and Hypothesis* in 1902, a wide range of different attempts to reconcile Newtonian mechanics and electromagnetic theory were still being attempted. Poincaré was substantially influenced by Kant and even more by two centuries of the Newtonian tradition. While he regarded large parts of Euclidean geometry and Newtonian mechanics as being conventions rather than truths, they were conventions which he thought would never be abandoned. In 1904 and 1905 respectively Poincaré and Einstein independently produced versions of the special theory of relativity. Poincaré shifted his ground, but he was not able to adapt to the philosophical implications of the new theory with the same ease as the much younger Einstein.

By dethroning the two best established aspects of classical science, namely Newtonian mechanics and Euclidean geometry, Einstein established himself as one of the towering geniuses of all time. He distinguished sharply between mathematics as an axiomatic subject, and its possible relevance to the physical world, writing:

> *Pure logical thinking cannot yield us any knowledge of the empirical world; all knowledge of reality starts from experience and ends in it. Propositions arrived at by purely logical means are completely empty as regards reality. Because Galileo saw this, and particularly because he drummed it into the scientific world, he is the father of modern physics—indeed, of modern science altogether.*[13]

Einstein's new insights were based on the proof by Riemann and others that there were many different geometries which were equally valid in purely mathematical terms. His reason for distinguishing between Euclidean geometry as mathematics and its physical relevance was not philosophical. His theory of relativity showed that the idea that the physical world is Euclidean is not only wrong in detail, but fundamentally misconceived. The same applied to the unquestioned belief of scientists that time was a straightforward concept about which everyone could agree:

> *At first sight it seems obvious to assume that a temporal arrangement of events exists which agrees with the temporal arrangement of the experiences. In general and unconsciously this was done, until sceptical doubts made themselves felt. For example the order of experiences in time obtained by*

acoustical means can differ from the temporal order gained visually, so that one cannot simply identify the time sequence of events with the time sequence of experiences.[14]

His idea that objects could only be described properly if space and time were considered together was already in the air at the time: H. G. Wells made it a principal point in the opening pages of *The Time Machine*, published in 1895. Out of such unpromising elements Einstein created a magnificent new theory in which time lost its absolute status and was amalgamated with space into a new entity called space-time. The reason for the importance of relativity theory lay in the fact that it made new predictions about certain phenomena. Although very difficult to measure in normal circumstances, the validity of the special theory of relativity for bodies moving at very high speeds is not a marginal issue. High energy accelerators such as the facility at Geneva have been producing billions of high energy particles over several decades. There is no doubt at all that their motion is in accordance with relativity theory and quite different from what Newton's laws predict.

Although Einstein's new theory superseded Newtonian mechanics, it turned out to be mathematically incompatible with the quantum theory which was to be invented in 1925/26 by Heisenberg, Schrödinger, and others. So by 1930 two brand new theories had come into being. Both demonstrated beyond doubt that Newton's theory was just that, a set of mathematical equations which provided extraordinarily good predictions in many situations but were not ultimately a *true description of reality*. Einstein worked for much of the rest of his life to find a way of combining relativity with quantum theory into a unified field theory, but he was not successful. His disbelief in the fundamental insights of quantum theory is now remembered with feelings of sadness. His claim that 'God does not play dice' is immortal, but almost all physicists now believe that he was wrong.

6.3 Discussion

Let us review what we have learned in the chapter. Galileo eventually triumphed over the authority of the Church because people could confirm his astronomical observations directly. For the Church to regard the Copernican theory merely as a convenient computational device, provided its physical truth was not advocated, was not a viable option by 1700. The Ptolemaic system faded into insignificance because it involved ever more elaboration as observations became more precise. Newton was undoubtedly a genius, but he was a genius who was born at the right time. His laws of motion provided an astonishingly good method of reconciling a wide range of physical phenomena using equations which did not need continual revision.

With the benefit of hindsight it is easy to ridicule the use of cycles and epicycles as hopelessly complicated, but in fact something extremely similar is in common use at the present time. The expansion of functions in Fourier series is based upon the same idea, that one should build up general periodic motions by a combination of circular motions of different amplitudes and frequencies. Of course Ptolemy had to explain his system in purely geometrical terms, while Fourier had the benefit of huge developments in algebra and calculus. The fact that planetary orbits can be described exactly in terms of ellipses was of tremendous importance at the time, because extended numerical calculations were extremely hard. Until late in the twentieth century science progressed by writing down solutions of problems in terms of a few well-known functions such as sine and cosine for which books of tables were compiled. No simple exact solution for the three body problem exists, but we are now undisturbed by this fact because computers enable us to handle such problems routinely.

We next come to the question of the correctness of the heliocentric theory. In spite of the great insights of Copernicus and Newton, we no longer believe that the Sun is stationary—we have a much wider view in which galaxies and even clusters of galaxies participate in a general expansion of the universe.[15] Ultimately we defer to the general theory of relativity, which tells us that all frames are equivalent, so that there is no meaning in asking whether the Sun or the Earth or indeed anything else is at the centre of the Universe. In spite of this, when contemplating space travel or designing satellite-based telephone systems, we use Newtonian mechanics within a heliocentric universe. Indeed we still talk in terms of the earlier geocentric theory when we refer to the Sun rising in the East every morning and setting in the West, because in many daily contexts that is the simplest language to use. The theory we use in most circumstances is not the most correct one but *the simplest one which fits the facts well enough.*

Certain consequences of the Galilean revolution are so fully incorporated into our way of thinking that it is impossible for us to imagine abandoning them. The Moon is now merely an object like any other, and the 'face' we appear to see is merely our interpretation of geographical (or rather selenographical) features on it. Similarly the Sun is a material object which radiates light and heat because of nuclear processes going on deep in its interior, which will eventually come to an end when the nuclear fuel is exhausted. The mystery of the Aurora Borealis is explained by the interaction of streams of charged particles emitted by the Sun with the Earth's magnetic fields. These purely materialistic explanations provide the context within which all of our thought takes place. Their explanatory power is so great and so consistent that it is inconceivable that we could abandon them, even if the details need working over on occasion.

There was, however, always a mystery at the centre of Newton's law of gravitation. Leibniz and others quite wrongly came to think that Newton believed in action at a distance. He provided no mechanism by which two remote bodies could attract each other, but it is clear from his private papers and correspondence that this was a matter of great importance to him, and that he did not feel satisfied with simply declaring what the laws were.[16] In the General Scholium Newton was, in fact, stating that the *truth* of the inverse square law should be separated from its *explanation*. He had demonstrated the former and made no public hypotheses about the latter. It is an interesting comment on the way in which physics has developed that current textbooks make no attempt to explain why gravity should obey an inverse square law by the production of a mechanical model. Indeed it seems that physicists now 'explain' the law by recourse to an even more abstract theory, namely general relativity in the approximation of an almost flat metric.

By the start of the nineteenth century Laplace and others were to claim that the application of mathematical laws would enable one to obtain the solution of any problem involving the motion of bodies, provided one had complete knowledge of their dispositions at some initial time. With the benefit of hindsight we can now see that basing a deterministic philosophy on the exact truth of Newton's laws was not a sound idea. One can often work out the orbit of a collection of planetary bodies accurately because there are only a small number of them, and they are essentially unaffected by the person applying Newton's laws to predict their motion. On the other hand there are situations in which the dynamics of the three-body problem is unpredictable because of its chaotic behaviour, and others in which the five-body problem literally has no solution beyond a certain time even though the bodies are involved in no collisions. The application of Newton's laws to very large collections of particles is even worse, in that chaos is generic rather than being dependent upon rather unusual initial conditions. If one fills a box with air molecules at room temperature and pressure, precise predictions of the motions of the molecules are impossible because of the limitations on the power of any computer. Even if computers could be sufficiently powerful, the degree of instability is such that the act of putting the data into the computer would disturb the molecules enough to render any computation worthless. The proposed calculation could only be performed by a computer which was simultaneously a part of the Universe and computing the effect of every part of the Universe on every other part. This is not possible.

John Polkinghorne has suggested that chaos theory encourages a belief that there are new causal principles at work in such situations, which have a holistic character. Newton's theory cannot be applied in sufficiently unstable situations, so something else may take over. There are several difficulties with this idea. The first is that there is no reason to believe that any new principle would bypass the extreme instability of the evolution of large assemblies of particles. The second is that holistic principles are very unlikely to have scientific content, because they do not provide the possibility of making precise predictions. One might be able to make statistical predictions in such situations, but it is well known that the same statistical conclusions can often arise from a variety of different detailed causes. On the other hand, if one *already has* good reasons to believe in the existence of some very subtle holistic principle, then chaos might explain why its influence is not more obvious.

My main point is that any mathematical description of a gas of particles is merely a model of reality, not reality itself. One cannot take any physical model to be relevant if its application involves referring to distances far smaller than the diameter of an atom. No physically relevant theory can necessitate computing to hundreds of digits, because we will never be able to measure anything that accurately. The use of real numbers in physical models of reality is not justifiable if one needs computations of that accuracy. If any physical theory seems to require this, then the theory has proven its own eventual failure, however impressive the experiments leading up to the theory were.

While I explained the errors in Laplace's claims by reference to the effects of chaos, historically speaking belief in the *truth* of Newton's theory was abandoned for other reasons. Einstein's general theory of relativity was discovered (or invented) in 1916. The prediction that light would be bent by intense gravitational fields was confirmed during an eclipse on 29 May 1919 by an expedition to Principe Island in the Gulf of Guinea funded by the Royal Society of London. In 1925–26 quantum mechanics provided yet another mathematical theory which yielded the same predictions as Newton's as far as the motions of everyday objects were concerned.

When Einstein developed the special theory of relativity he abandoned the 'self-evident' idea that space and time were different types of entity, and in general relativity he also gave up the idea that space-time was flat or structureless. However he still clung, as subsequent physical theories have, to the ultimate continuity of the universe. On the other hand Richard Feynman, a Nobel prize winner for his contribution to quantum electrodynamics, speculated in his Feynman lectures that the ultimate structure of space-time might be quite different from that conceived by Einstein. In the last twenty years we have had theories presented involving it being ten or eleven dimensional, or having a very complicated topological structure at the smallest level. In some research papers the 'extra' dimensions are supposed to be non-commuting. Current fundamental physics indicates that one cannot measure anything to a finer resolution than a certain very small distance, the Planck length of about 10^{-35} metres, and a very short time, the Planck time of about 10^{-43} seconds. It is entirely possible that the mathematics which we have developed to describe the world may be wholly inappropriate at a fine enough level, and that some day a model will be devised in which space and time are actually discrete.

In the end we have to remember that the Universe is an entity, not a set of equations. We can try to isolate some part of it and predict the behaviour of that part using mathematics, or by other means. The success of physics consists of making mathematical models which are simple enough for us to be able to solve and yet complicated enough to capture some interesting aspect of the world. A precondition for being able to carry out calculations is that the aspect of the world being studied is closed (unaffected by the outside world) to a good approximation, or that the influence of the outside world can be summarized in a sufficiently simple manner. Our theories and methods of analysis work extraordinarily well in a huge variety of such simple situations, but that does not mean that we can assert that they *would* apply to the whole Universe *if only* the relevant computations could be done, when we know that they cannot. Nor do we have the right to claim that because a theory provides wonderfully accurate predictions in a variety of special situations, it must be *true* in some deep philosophical sense.

Notes and References

[1] By the Church I shall always mean the Roman Church, without suggesting that the Lutheran or Calvinist churches were any more tolerant of dissent. They were not!

[2] The manufacture of glass was one of the key technologies underlying the scientific revolution of the seventeenth century, and was one of the very few technologies of that period which did not originate in China.

[3] Ariew 2001

[4] The full title is much longer.

[5] Galileo 1632, p. 108

[6] Actually he proved that planetary orbits were elliptical *in the two-body approximation*, and acknowledged that the influence of Jupiter would lead to discrepancies from strict ellipticity.

[7] This had been awarded its royal charter in 1662 by Charles II.

[8] This is taken from the opening section 'Rules for the Study of Natural Philosophy' of Book Three. I follow the recent translation of *Principia* by Cohen and Whitman, which also contains many comments about recent scholarship on Newton. [Cohen and Whitman 1999]

[9] Laplace was well aware that the calculations would have to include the effects of electricity and magnetism, not then at all well understood, as well as of gravity.

[10] The eccentricity measures the extent to which the orbit deviates from a circle, and equals 0.1236 for Hyperion.

[11] Black et al. 1995, Murray 1998, Murray and Dermott 1999

[12] Saari and Xia 1995

[13] Einstein 1982b

[14] Einstein 1982c

[15] One of Newton's very few mistakes in *Principia* was to state that the centre of the Solar System was in a state of absolute rest. This was a philosophical principle, for which he provided no convincing evidence.

[16] Cohen and Whitman 1999, ch. 9

7
Probability and Quantum Theory

7.1 The Theory of Probability

Probability is a very strange subject. The Encyclopaedia Britannica has two separate articles on it, one mathematical and the other philosophical. The first gives the impression that the theory is completely straightforward while the second states that the proper interpretation is a matter of serious controversy. These controversies are still being actively discussed.[1]

One standard interpretation is that probabilities represent the frequency of occurrence of events if they are repeated a large number of times. Thus one may justify saying that the probability of getting a head when tossing a coin is a half by tossing it a sufficiently large number of times, and observing that one gets a head on about a half of the occasions.

A straightforward frequency interpretation of probabilities is inappropriate if one asks for the probability that a hot air balloon will make a forced landing in Dulwich Park, South London, next 13 June. The problem is that no such event has occurred in the past, so one has to look at the frequency of *similar* events. What counts as similar is, unfortunately, a matter of judgement. Thus one might decide to find out the number of times at which hot air balloons have landed in any London park, sports ground or golf course on any day in May, June, or July in the past. If this number is reasonably large, the required probability can be inferred, subject to certain assumptions. To name just one, Dulwich Park is very close to three other large open areas, so a balloon might be less likely to make a forced landing there than if it were more isolated. We conclude that even in this case, the frequency interpretation would have to be combined with considerations which are far from elementary.

Laplace regularly used probability theory in situations in which a frequency interpretation was out of the question. As an example, he carried out a calculation of the mass of Saturn, concluding that there were odds of 11000:1 on the mass of Saturn being within 1% of a certain calculated value. Now there is only one Saturn and its mass is not a random quantity. Laplace was using probability theory to express confidence in a calculated value, based on prior beliefs as modified by subsequent evidence and calculations. As statisticians have had to cope with ever more complex problems over the last thirty years, they have realized that *prior judgements* about what is relevant to a particular problem are an essential part of the subject. Bayes' theorem tells one how to update one's *prior beliefs* in the light of subsequent observations. In complex situations it is simply not possible to carry out some standard 'objective' analysis of the data which leads directly to the 'right' answer. Ed Jaynes put it as follows:

> *In Bayesian parameter estimation, both the prior and posterior distributions represent, not any measurable property of the parameter, but only our state of knowledge about it. The width of the distribution is not intended to indicate the range of variability of the true values of the parameter ... It indicates the range of values that are consistent with our prior information and data, and which honesty therefore compels us to admit as possible values.*[2]

In this chapter we will discuss a series of examples which illustrate various peculiarities of probability theory. Some of these are included simply for their entertainment value, but I also have a serious purpose. This is to establish that there are many situations in which two people may legitimately ascribe different probabilities to the same event. These observer-dependent aspects are important when the two people have differing information about the events being observed. A complete separation between observer and observation may be possible in Newtonian mechanics, but it does not always work in probability theory, or in quantum theory as we shall see later. This is not to say that probability or quantum theory are purely subjective—used properly they make definite predictions which *work* in the real world.

Kolmogorov's Axioms

Probability theory arose from attempts to devise strategies for gambling in the seventeenth century. The single most important result in the field, the central limit theorem, is due to Laplace in 1812. In 1827 the botanist Robert Brown observed that minute particles of pollen in water could be observed under a microscope to move continuously and randomly. Einstein's quantitative explanation of this phenomenon in term of the random buffeting of visible particles in a liquid by the molecules of the liquid became one of the key proofs of the atomic theory in 1905, as well as establishing the fundamental role of probability in physics.

Since the motions of pollen grains depend upon molecular collisions, these also determine whether a particular pollen grain is eaten by some other organism or survives to produce a new plant. We therefore see that real events at our own scale of size may be unpredictable because they depend upon chaotic events occurring at the molecular level.

Einstein's paper made further points of general interest. Small enough visible particles subject to random impacts with molecules in a liquid could not be said to have an instantaneous speed. Indeed the average distance they moved in a short period would be proportional to the square root of the time elapsed, rather than to the time itself. This would have the consequence that attempts to determine their speed would give larger and larger values the shorter the time interval that was considered. This immediately explained the failure of all previous attempts to measure just this quantity![3] It would later be incorporated into a complete theory of stochastic processes.

The modern era of probability theory dates from 1933, when Kolmogorov formulated it as an axiomatic theory. Probabilists' subsequent concentration upon the sample path analysis of stochastic processes is entirely set in this context. Kolmogorov's axioms enabled mathematicians to cut themselves free of the philosophical controversies surrounding the subject, and to concentrate upon what they did best. As time passed many eventually came to believe that probability theory and Kolmogorov's formulation of the subject were indistinguishable: that there was no other possible coherent account of the subject.

Kolmogorov's theory may be summarized as follows. One first has to define the possible outcomes of the experiment or problem. For three tosses of a coin the possible outcomes are HHH, HHT, HTH, HTT, THH, THT, TTH, TTT. If we are grading oranges into sizes by measuring their diameters, then each orange is associated with a number in the range 0 to 10 (measured in centimetres). The result of grading a dozen oranges is a sequence of 12 such numbers. Each such sequence is a possible outcome.

The next step in the theory is to associate a probability with each possible outcome, in such a way that the sum of all the probabilities equals one.[4] These probabilities are assigned after a careful consideration of the particular problem. For example if you toss a fair coin 3 times then each of the outcomes listed above would have probability $\frac{1}{8}$; this is essentially what is meant by saying that the coin is fair. At the other extreme if you knew that you had a double-headed coin then you would assign HHH the probability 1 with all the other probabilities equal to 0. A more interesting situation arises if you have two coins in your pocket, one of which is fair while the other is double-headed. On taking out one without looking at it and tossing it three times, you should assign HHH the probability $\frac{9}{16}$ while all the other outcomes have probability $\frac{1}{16}$.

The total number of possible outcomes is often very large, and each outcome may have an extremely small probability: thus any *particular* sequence of heads and tails in a long succession of tosses of a coin is extremely unlikely. Because of this one is often more interested in probabilities of collections of outcomes. Probabilists use the term 'event' to denote a collection of outcomes with some common property. Thus getting two heads in three tosses of a fair coin corresponds to choosing the event

$$TwoHeads = \{HHT, HTH, THH\}.$$

Each individual outcome has equal probability $\frac{1}{8}$, so the probability of the event TwoHeads is

$$\text{Prob}(TwoHeads) = \tfrac{1}{8} + \tfrac{1}{8} + \tfrac{1}{8} = \tfrac{3}{8} = 37.5\%.$$

Kolmogorov's theory thus involves three elements: selecting the 'sample space' of possible outcomes, choosing the probability law which is most appropriate, and devising procedures for calculating the probabilities of quite complicated aggregates of the individual outcomes. To a first approximation probabilists assume that the first two stages are given and concentrate on the last. Statisticians examine data and use a variety of statistical methods to select the most likely probability law out of a range of previously chosen possibilities.

Although the rules of probability theory are straightforward, they lead to more paradoxes in applications than any other subject apart from quantum theory, which is also probabilistic. I will describe a few and explain how they should be resolved.

Disaster Planning

One of the problems facing governments is deciding whether to allocate resources to prevent or cope with disasters which have never happened, and whose likelihood is very unclear. Such decisions must be based on a trade-off between estimates of the likelihood of the disasters, the damage done if it occurs, and the cost of taking precautions against it. Statisticians have to fight a constant battle to stop politicians concealing facts which the latter find inconvenient.

Unfortunately estimates of the likelihood of many disasters depend upon assumptions whose accuracy may never be known. Consider the year 2000 computer bug. Before the event there were many predictions of the dire consequences of taking no action. It was suggested that if even one major bank was not able to process its transactions, this might have rapidly escalating consequences, leading even to the collapse of the world banking system. In the end, of course, nothing happened, but was this because the horror stories led major institutions to take measures which they might otherwise not have?

There are cases in which experts have been spectacularly wrong: namely in the calculation of the risk of major nuclear disasters. I well remember reassurances that modern management techniques would reduce the risk to about one serious event every million years. Sad to say there have already been three—at Windscale, Three Mile Island, and Chernobyl, each in a different country. In one rather technical sense the calculations of the experts were not wrong; the problem was that they did not consider the *effects of inactivity* on people who had to manage systems for long periods of time, during which nothing giving the appearance of danger ever happened. When no serious problems arose at the operational level, people eventually convinced themselves that the precautions were not necessary. The same happens when people drive cars, but in that case blunders only cause a small number of deaths.

At the present time we face an even more frightening scenario: that of the deliberate release of a highly infectious organism by terrorists. This presents an extreme case of each of the three problems mentioned above. Effective prevention would be extremely expensive, and would also result in a substantial loss of accustomed democratic freedoms. The cost of such an event might be measured in millions or even hundreds of millions of deaths. It was traditionally considered that nobody would be willing to carry out such an attack because of the immensity of the consequences, but that argument no longer appears persuasive. The use of probabilistic risk calculations in such a context seems to the author to be wholly inappropriate.

I have long been surprised that a *natural* pandemic has not yet occurred. Aids is close to this, but even worse possibilities can be imagined. Imagine a mutation of the wind-transmitted foot and mouth disease which infects human beings as well as sheep, and is usually fatal. With air travel at its present level, this could have spread to every country in the world before its existence was even known. Perhaps the only measure we can take to prevent it is banning most international travel, but who will advocate this in the absence of any proof of necessity?

The Paradox of the Children

Let us turn away from such morbid fears, and discuss a lighter topic. A woman with two children meets a stranger at the funeral of her aunt and the following conversation ensues.

> W. *I will never wear my aunt's diamond ring, but it seems a shame to sell it.*
>
> S. *If you have a daughter you could keep it for her when she grows up.*
>
> W. *What a good idea! I will do that.*

Based upon this information, what is the probability that the woman has a son? One approach, which leads to the wrong answer, is the following. You can infer that the woman has a daughter from her final statement. Since you know nothing about the other child, it is (more or less) equally likely to be a boy or a girl, so the answer is 50%. The correct approach uses conditional probabilities—which in this case means just keeping careful track of all the of possibilities. Before the conversation there are four equally likely possibilities BB, BG, GB, GG for the woman's family, where the first letter refers to the gender of the elder child, and the second to the gender of the younger child. The conversation reveals *only* that the combination BB is not possible, so there remain three equally likely possibilities BG, GB, GG. Two of these involve a son, so the correct probability is $\frac{2}{3}$, $\sim 67\%$.

The point of this paradox is that one must be extremely careful about what information is provided when computing probabilities. If the woman had said that her elder child was a girl, the answer would indeed be 50%. Even professional statisticians sometimes make mistakes by confusing two situations which differ only in such details.

The Letter Paradox

A person has two cards, one with the word HEADS written on it, and the other marked TAILS. He puts the cards into two identical envelopes and shuffles them so that he does not know which card is in which envelope. He then sends one letter to a friend Belinda in Belgium, B, and the other to a friend Charles in Canada, C.

Clearly the probability that C has the HEADS card is 50%. B now opens her envelope to find that it contains the TAILS envelope. So the probability that C has the HEADS card suddenly changes to 100%, even though C does not know this. At this point B opens a parcel which contains a bomb, killing her and destroying the card. Although tragic for B, the probability that C has the HEADS card must surely remain 100%. Now suppose that B had opened the parcel first, killing her and destroying the unopened letter. Would this change anything?

The paradox is resolved by accepting that the probabilities are not attached to the cards alone. In other words the probabilities describe the ignorance of particular people about the cards.[5] For C the probability of getting the HEADS card remains at 50% whatever happens to B. The latter, or someone who knows what the sender actually did, would correctly assign a different probability to the event. I emphasize:

> Two different people can correctly assign different probabilities to the same event, reflecting their different degrees of ignorance about the true situation.

This is an important idea in the interpretation of probability theory. Note the mention of the 'true situation'. For a person fully informed about this *there are no probabilities*.

The Three Door Paradox

In a television game show the studio has three doors behind one of which is a prize. The contestant chooses one of these but it is not opened. The host, who knows where the prize is, opens a different door to reveal that there is nothing behind it. Should the contestant change his/her mind? Here are two arguments.

(1) The prize is behind one of the remaining two doors and is as likely to be behind the one already chosen as behind the only other one left, so there is no reason to change one's mind.

(2) Originally the chance of winning was $\frac{1}{3}$, but in the new situation there are two choices left, so the best thing is to make a new independent random choice between those two.

Both of the above are wrong. It is widely agreed by probabilists that the correct strategy is to change one's mind and choose the other remaining door! Their reasoning is as follows. The possible outcomes are 11, 12, 13, 21, 22, 23, 31, 32, 33, where the first digit refers to the door where the prize is and the second refers to the choice of the contestant. Each of these has the same probability $\frac{1}{9}$. The host now has to open a door with nothing behind it. In the cases 11, 22, 33 the contestant should not change his mind, but in the cases 12, 13, 21, 23, 31, 32 he should. So the probability that it is better to change is $\frac{2}{3}$.

However, even this argument can be criticized, because it assumes that all of the original nine possibilities was equally likely. In fact people have very strong unconscious biases when making 'random' choices. If the contestant knows that the host, being human, is likely to open the middle door whenever that is possible within the rules, or that the prize is more likely to be behind the middle door than any other, this should influence the way he plays. Lest you think this is far fetched, I should mention that stage magicians regularly exploit people's failure to make even approximately random choices. When I recently asked a group of 25 students to choose a number at random, almost all of them chose a number under 11, none of them chose 1 or 2 and almost a half chose 7 or 8.

The National Lottery

When the British National Lottery started, there was one event per week, which involved the random selection of six balls from a larger set numbered 1 to 49 inside a rotating drum. Within a few months people were collecting statistics avidly, hoping to find which numbers were lucky or unlucky. There were soon enough complaints about unfairness for the Royal Statistical Society to be asked to provide advice about them. Needless to say the effects noticed were the results of people's misconceptions about the behaviour of random trials rather than problems with the Lottery.

I conducted a numerical experiment, selecting numbers randomly from 1 to 49 with replacement: this is not quite appropriate to the Lottery in which the same number cannot occur twice in the same week. I stopped after 182 trials (7 per week for 26 weeks). At this point I listed the frequency with which every number had occurred in order starting from 1 and ending with 49:

$$7, 2, 4, 2, 3, 3, 5, 4, 5, 6, 3, 3,$$
$$5, 2, 3, 2, 5, 7, 4, 8, 2, 1, 1, 6,$$
$$2, 3, 2, 5, 4, 3, 6, 5, 7, 4, 2, 4,$$
$$4, 1, 3, 5, 2, 2, 2, 2, 4, 5, 5, 2, 5$$

The most unlucky numbers 22, 23, 38 occurred only once while the luckiest number, 20, occurred 8 times. Similar results were found in other trials, except that the particular numbers which were most or least lucky changed on each occasion. In a dozen trials the 'worst' had one number never occurring and another occurring 11 times.

These results are not surprising to a statistician, but they are exactly what thousands of people have been studying in the belief that they will help them to win the lottery. The lesson is that people expect that the results of a random trial will be considerably closer to the 'exactly equal' outcome than either the theory or the facts warrant.

Although nothing can help one to win, there IS a strategy to increase your profit in the unlikely event that you do win. Many people bet using their birthdays, which must be less than 32, or numbers which they have good feelings about. These good feelings may be related to their appearing in multiplication tables and so being 'familiar'. This suggests one uses large primes, in order to reduce the chance that someone else has made the same choice as you. So a good combination is

$$47, 43, 41, 37, 31, 29, 23$$

Unfortunately I just made the likelihood of people choosing this combination increase by publishing it. This is an example of the 'reflexivity' which bedevils the social sciences, that a successful theory of how people behave may immediately be used by some of those very people to their own advantage. Such behaviour renders the theory no longer valid. Perhaps the only laws in the social sciences are those which are kept secret by those who discover and then exploit them!

Probabilistic Proofs

Suppose that one is asked to prove an identity such as

$$(x^2 - 3x + 3)^4 = x^8 - 12x^7 + 66x^6 - 215x^5 + 449x^4$$
$$- 613x^3 + 544x^2 - 300x + 81.$$

One could do this simply by expanding the left-hand side, but it would be fairly painful, unless you have access to a computer algebra package. A second idea is to check the identity for $x = 0, 1, 2, 3, 4$. It is true in all of these cases, so the temptation is to say that it is probably true (actually it is not).

There is a quite different and very simple method of demonstrating the validity of more or less any formula involving only one variable (I use the word demonstrate, rather than prove, deliberately). This is to choose a real number x at random, and check the identity for that single number. If the formula is true to the usual accuracy of a computer or pocket calculator for that value of x, then it is almost certainly true for all x. Or to put it the other way around, if the formula is not true, then you would be incredibly unlucky if it turned out to hold for a randomly chosen value of x.

There is actually a potentially rigorous way of proving polynomial identities along the above lines. If the polynomial has integer coefficients and the identity is true for the single value $x = \pi$ then it is true for all x. This remarkable property of the number π is of less practical use that might be thought: one cannot evaluate the required polynomial *exactly* because computations always involve rounding errors, and π has infinitely many digits. These vary in a highly irregular manner. Indeed they satisfy all known tests of randomness, although they are not random in any true sense.

The following type of empirical demonstration is close to one used by Euler in the eighteenth century and discussed on page 114. Suppose that you have produced the tentative identity

$$\frac{945}{\pi^6} \sum_{n=1}^{\infty} \frac{1}{n^6} = 1$$

by an abstract method in which you have little confidence. In order to check whether it is plausible you might then evaluate the quantity on the LHS but only adding up the first three terms of the series. One then gets the number 0.9996595 ..., which is close enough to demonstrate that the identity is worth further investigation. At this point one can make the testable hypothesis that if one evaluates the LHS but adding up the first fifty terms then one should get a much closer approximation to 1. The result of this second computation is the number 0.9999999994 .... There is only one chance in a million that the first six *incorrect* digits in the first approximation should become correct in the second if the identity is not true.

The following story will serve to explain why the first calculation should not be accepted as it stands, in spite of the closeness of the value obtained. About a year ago I was in the Louvre in Paris, and decided to retrace my route by a few rooms to see something which I had missed. On the way I bumped into another mathematician whom I had not seen for several years. What an amazing coincidence—or was it? As we walk around we constantly scan the people near us, even if only to avoid walking into them. On that day alone I had probably glanced at a few thousand people, and if I count since the previous chance meeting there were probably hundreds of thousands of other non-coincidences—people I passed whom I did not know. We humans have an astonishing tendency to attach significance to random events, probably because to fail to recognize a pattern might have more serious consequences than to 'find' patterns where there are none. The moral is that uncontrolled observations should always be regarded as no more than a source of hypotheses which can then be tested scientifically.

Stopping trials as soon an unexpected result occurs, in order to tell everyone about your amazing discovery, and forgetting all the coincidences which did not happen, is extremely common. It is responsible for the so-called Torah codes and for claims about ESP.[6] Drug companies have to make major efforts to avoid being drawn into the same trap.

What is a Random Number?

It is commonly claimed that if one chooses a number from 1 to 10 at random, then each of them should be assigned the probability $\frac{1}{10}$, unless some other information is given. Unfortunately there is no such soft option. One cannot do probabilistic calculations without considering the way in which the probabilities arise. The fact that equal probabilities are relevant to tossing 'fair' coins or rolling dice involves a judgement about how those actions are performed physically.

If one tosses two dice, each numbered from 1 to 6, and then adds the scores, one gets a total score between 2 and 12. Assuming that the dice are both fair, the probabilities of each of the scores is as follows:

Score	2	3	4	5	6	7	8	9	10	11	12
Probability	$\frac{1}{36}$	$\frac{2}{36}$	$\frac{3}{36}$	$\frac{4}{36}$	$\frac{5}{36}$	$\frac{6}{36}$	$\frac{5}{36}$	$\frac{4}{36}$	$\frac{3}{36}$	$\frac{2}{36}$	$\frac{1}{36}$

If you do not know how the numbers from 2 to 12 were generated, and start betting on the basis that they have equal probabilities, then you will lose a lot of money until the data lead you to a better understanding! Of course, if you start with a sufficiently large data set then such problems do not arise, but real life is seldom like that.

Let us count the lengths of words in an article, omitting all words of length greater than ten. The data in figure 7.1 were taken from the Economist magazine.

While longer words are less common, we also see that words of length 1 are not very common. Although an experimental fact, the above probabilities are highly dependent upon the context. If you pick up a different magazine, read an article written by a different person or even by the same person when he or she has a headache, you may find a different distribution of word lengths. Your predictions should be affected by all of these bits of information. The more you know about where the passage comes from, the better your predictions are likely to be. There is no a priori 'best guess' in this situation: if you have no

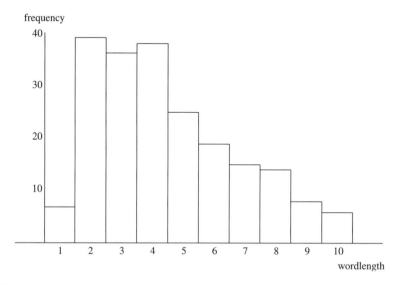

Fig. 7.1 Wordlength Frequencies

idea of the correct distribution and assume that every word length from 1 to 10 is equally likely then your predictions will be extremely poor, until your failure forces you to change your mind. The moral is that if you do not know where some 'random' numbers come from then you should spend your time thinking about what is likely to be their distribution, rather than betting in ignorance.

Bubbles and Foams

Foams, defined as materials densely packed with tiny spaces, are ubiquitous. Sponges and the interiors of bones are highly porous, but cork and expanded polystyrene are composed of separate cells, and are therefore good insulators. Metal foams are becoming increasingly important as a source of lightweight materials with novel properties. For such purposes, it is desirable to make the size of the cells as uniform as possible. In other contexts their sizes can vary enormously.

Let us consider a bottle containing a little soapy water. Shake it until it is full of bubbles and then leave the bubbles to settle down. Gradually more and more of them collapse until there are a few very big bubbles left in the bottle, as well as a lot of smaller ones. The very schematic figure 7.2 represents a two-dimensional section through the bubbles. We ask what size a randomly chosen bubble is likely to be.

The obvious method is to calculate the average (or possibly the median) size of the bubbles and declare that to be the answer. This idea has few merits, and there is in fact no answer to the question posed. Let us imagine that there are many bacteria floating in the air in the bottle. One way of picking a bubble is to choose one of the bacteria 'randomly' and then to ask which bubble it is in. This method of choice is most likely to pick one of the larger bubbles, because

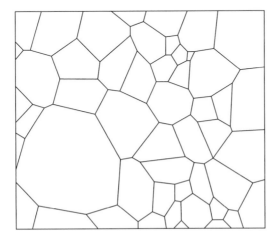

Fig. 7.2 Soap Bubbles

the number of bacteria inside a particular bubble is likely to be proportional to its volume. In figure 7.2 the largest bubble has volume a few thousand times that of the smallest, and so is far likelier to be chosen by this method. The moral is that until one knows the mechanism by which the choices are made, it is meaningless to talk about probabilities.

The above argument is clearly uncontroversial once one has thought about it, but a fallacious discussion of a somewhat similar problem (the 'design' of the universe, discussed in Chapter 10) is widespread. Imagine an intelligent bacterium, which sees that the vast majority of the bubbles in the bottle are much smaller than the one it is in. It might argue that this was so improbable on the basis of pure chance that it must have been placed in one of the biggest bubbles by design, but it would be wrong. The *only* justification for assigning equal probabilities to various events is that, after careful consideration, one can see no reason for believing any of them to be more likely than any other.

Kolmogorov Complexity

In the above examples the randomness observed is not inherent in events themselves, but is dependent on our ignorance of relevant information about them. Here we discuss an attempt by Kolmogorov and others to give an objective meaning to the phrase 'a random sequence'.

If one has a long string of digits, one can consider the shortest program which will generate that number in a particular programming language. Clearly if the number x has 1000 digits then one such program would be 'Print(x)' which involves 1007 symbols. But many numbers have other, much shorter descriptions. For example the number $10^{1000} - 1$ may be described by 'Print(9) 1000 times'. which involves only 19 symbols, although the number itself has 1000 digits. Although some numbers have shortest descriptions much smaller than their number of digits, a simple counting argument shows that numbers with very short descriptions are very uncommon. Kolmogorov suggested that one should define a random string of digits as one which does not have any description radically shorter than the length of the string. It may be shown that this notion does not depend essentially on the programming language used in the definition.

For each string of digits (i.e. number) x there exists a number $f(x)$, called the Kolmogorov complexity of x, which gives the length of the shortest procedure for generating that string. We have already shown that $f(x)$ is certainly no bigger than $n+7$ where n is the number of digits of x. Unfortunately at this point the subject starts to fall apart. One method of evaluating $f(x)$ is to examine all programs with length up to $n + 7$, see which of them generates the given string and determine the shortest of those. Unfortunately this is wholly impractical, and it has been shown that there is no short cut. There does not exist a systematic procedure for determining the minimum length program which improves substantially upon the brute force search. A proof of this is beyond the scope of the present book, but the following example makes it plausible. The digits of π are perfectly random according to a wide series of statistical tests as far as they have been computed (about the first trillion are known), but a quite short description of how to generate them is obviously possible. As a result if one were to write down the digits of π starting not with the first but with the millionth, it would take great effort or considerable luck to discover the non-randomness of the sequence.

The above is an example of a subject which is completely conventional in mathematical terms, but which has fallen from favour because it is uncomputable. Contrary to popular opinion, fashions do affect what mathematicians study. The move towards topics which have a computational aspect seems sure to continue for a considerable time. Mathematical fashions last so long, decades at least, that they often appear to be permanent aspects of the subject, until they are overwhelmed by the next fashion. But previous fashions are preserved in folk memory, often to be brought out and dusted off ages later.

7.2 Quantum Theory

Quantum theory is perhaps the most difficult field of science to explain to a general audience. The mathematics involved is highly sophisticated and the subject is full of paradoxes which still resist *intuitive* understanding, even by experts. On the other hand, *technically* quantum theory is very well understood. In a wide variety of experimental situations involving atomic interactions physicists know the precise mathematical equations which govern the motions of the particles concerned. The rules for setting up and solving the equations are not controversial. They yield predictions about the behaviour of microscopic systems which are regularly confirmed to high accuracy in laboratory experiments. Even when an experiment is deliberately designed to test a particularly bizarre prediction, the phenomenon regularly appears as the theory predicts.

Several attitudes towards the interpretation of the formalism have emerged. The most cautious is that one should not ask such questions, because science is about making correct predictions and not about providing mental pictures as psychological crutches. Another is that quantum particles are strange entities, partly particles and partly waves, and that we have to accept that we are not mentally equipped to understand their essential nature. Yet another view is that we must continue to struggle to find the correct interpretation, which will remove all of the paradoxes once we have it.

In all of the above it is accepted that the equations of quantum theory are correct. Some people believe that quantum theory must be incorrect, and that future discoveries will replace it by something more accurate and simultaneously more easily comprehensible. One suggestion is to include irreversible terms in the evolution equation which have not so far been noticed because of their very small size. Unfortunately all such proposals have one of three defects: they have serious structural problems, have no experimental evidence to support them, or are so vague as to be mere aspirations.

In this half of the chapter we discuss a few of the well-known paradoxes of quantum mechanics. I will press one particular interpretation of the subject, which insists upon a role for the observer, but not for his/her consciousness. This interpretation is far from being my own invention, and strikes me as being more convincing than any of the others currently advocated. It will not resolve the deepest mysteries of the subject, because this is too much to ask at the present time.

History of Atomic Theory

Democritus was the first person we know to have advocated an atomic theory of matter, in 430 BC. Lucretius' first century BC poem *The Nature of the Universe* explained Democritus' ideas, which were astonishingly accurate for the time:

> On the other hand things are not hemmed in by the pressure of solid bodies in a tight mass. Thus is because there is vacuity in things. ... by vacuity I mean intangible and empty space. If it did not exist things could not move at all. For the distinctive action of matter, which is counteraction and obstruction, would be in force always and everywhere. ... Material objects are of two kinds, atoms and compounds of atoms. The atoms themselves cannot be swamped by any law, for they are preserved by their absolute solidity. ... The number of different forms of atoms is finite.

However, Lucretius also made statements which we now regard as wholly misguided. The Greeks and Romans had no means of proving or disproving the atomic hypothesis, which was made on philosophical rather than scientific grounds. Seventeenth century scientists such as Boyle, Hooke, and Halley had a lively interest in atomic theory, but an experimental proof of their existence was still far beyond them.

A landmark in the development of scientific chemistry was Lavoisier's *Traité élémentaire de chimie*, published in 1789, five years before he died under the guillotine. This gave a correct chemical account of combustion, established a systematic notation for acids, bases, and salts, and provided the first true table of the chemical elements. It might be described as the chemists' equivalent of Newton's *Principia*. In 1808–10 John Dalton's two volume *New System of Chemical Philosophy* described precise quantitative laws for the combination of elements. Dalton also strongly suggested that matter must be atomic. He wrote:

> It is one great object of this work, to shew the importance and advantage of ascertaining the relative weights of the ultimate particles, both of simple and compound bodies, the number of simple particles which constitute compound particles.

Dalton used his results on the relative masses of the atoms of different elements to describe the structures of many chemical compounds. By these means he was able to explain a wide variety of different reactions in a systematic manner.

Dalton's theory was widely recognized, but was not regarded as definitive proof that matter was composed of atoms for many years. There were good reasons for this: the actual structure of quite simple molecules was not easy to settle, and in fact many of the atomic weights determined by Dalton were incorrect. This was partly because he assumed that water molecules contained one atom each of hydrogen and oxygen, rather than two of hydrogen and one of oxygen. Although Avogadro resolved many of the confusions of the subject in 1811, his work was not recognized for another fifty years. It was only in 1869 that Mendeleyev produced a reasonably complete periodic table of the elements.

Dalton's geometric representation of molecules was the first attempt at structural chemistry, but it was largely guesswork and was generally rejected in a British Association meeting in 1835 in favour of the algebraic notation of Berzelius. Although flawed in its details, Dalton's belief that determining the geometrical shape of molecules would be central to the understanding of isomerism and other aspects of chemistry was to be triumphantly vindicated in the twentieth century. Of course accepting that the shapes of molecules had an important influence on their chemical properties made no sense unless one also agreed that molecules were real objects, which many nineteenth century chemists did not.

In 1814 Fraunhofer viewed the light from the Sun through a spectroscope, and discovered a large number of sharp dark 'spectral' lines. These were classified over the rest of the century and identified with the spectral lines of individual elements, which could be observed by heating them in a flame in the laboratory. One set of lines could not be found on Earth, and in 1868 these were associated with an unknown element named helium, after the Greek word helios for Sun. The element helium was eventually extracted from the mineral cleveite in 1895.

Nineteenth century scientists had no means of observing individual atoms because of their tiny size, and many regarded them as no more than a simple, and therefore convenient, way of summarizing experimental results. Leading among these was Ernst Mach, who wrote the following as late as 1896:

> *The* heuristic *and* didactic *value of atomistics* ... *should certainly not be denied. It is significant that Dalton, who was a schoolmaster by trade, revived atomistics. But atomistics, with its childish and superfluous accompanying pictures, stands in sharp contrast to the other philosophical developments of modern physics.*[7]

Within twenty years such views had been abandoned by all reputable physicists. The 'schoolmaster' had the last laugh!

The final evidence for the reality of the basic particles of matter came at the turn of the century. Electrons were discovered by Thompson in 1897 and were initially known as cathode rays, because of the manner of production. (We have to pass over many others who made important contributions to the study of their properties.) Millikan was the first to 'see' individual electrons in very ingenious experiments. These measured the movement of a tiny oil drop suspended in the air when it carried a single surplus electron, making it slightly charged. In 1911

Rutherford observed the anomalous scattering of alpha particles, which he could only explain on the basis that the atoms of matter were composed of even tinier nuclei around which electrons were orbiting.

Although there was no explanation of the spectral lines of atoms in terms of Newtonian mechanics, it was recognized that they had some connection with the way in which electrons orbited around the atomic nuclei. Unfortunately, classical electromagnetic theory predicted that such orbiting electrons would have to emit radiation, losing energy in the process and consequently spiralling in towards the nuclei. So even the existence of stable electronic orbits was a mystery. Many attempts to find a new theory were made, the most impressive of which was by Bohr. In 1913 he proposed quantization rules, according to which only certain electron orbits were physically permitted, so they could not decay. His theory correctly predicted the spectral lines of hydrogen and the helium ion, but failed for the helium atom. It was recognized as being ad hoc, and the search for a more complete theory continued.

The final breakthrough came in 1925 and 1926. In June 1925 Heisenberg invented a new matrix mechanics, in which Newton's notions of position and momentum were generalized by replacing real numbers by matrices, thus allowing an entirely new type of mathematics to be used in atomic theory. Between January and March 1926 Schrödinger invented a quite different wave mechanics, based upon the use of the spectral theory of partial differential equations. He then proved that his theory was essentially equivalent to that of Heisenberg. There were of course many other people involved, but by the end of 1926 the new theory was in place and many people were actively engaged in working out its implications. Most people's doubts about its correctness were laid to rest when the energy levels of helium were correctly calculated. This was a numerical computation done twenty years before the invention of computing machines, when the word 'computer' meant a person employed to do long series of calculations by hand.

The Key Enigma

In *The Feynman Lectures on Physics* Richard Feynman described a simple phenomenon which is absolutely impossible to explain in any classical way, and which is at the heart of quantum mechanics. He claimed, indeed, that it contains the *only* mystery of quantum mechanics. Because he was writing for physics students in Caltech, I will adapt his account.

Our starting point will be the photoelectric effect. Light striking a metal surface causes the emission of electrons. It is observed that more intense light does not increase the energy of each electron emitted, but does increase the electric current, that is the rate of emission of electrons. On the other hand the higher the frequency of the light (that is the closer the colour is to the blue end of the spectrum) the more energetic are the emitted electrons. Einstein gave a precise formula relating the relationship between the energy of the electrons and the frequency of the incident light, but this need not concern us here. More importantly his explanation of the effect established the particle nature of light as a fact.[8]

Einstein's idea was that light is composed of particles called photons. More intense light is simply light which has a greater number of photons passing each second. Each photon knocks one electron out of the metal, so an increased intensity of the light serves to increase the rate at which electrons 'boil' off the metal, but does not change the character of the electrons themselves. On the other hand the frequency of the light is directly related to the energy of each individual photon, more energetic photons having a higher frequency. The photons transfer their energy to the electrons which they knock out of the metal, so higher frequency light gives rise to electrons of higher energies.

This insight seems to be decisive, but in other situations the particle interpretation appears to be untenable. One of these is the so-called double slit experiment. This has been carried out for photons, electrons, and atoms such as rubidium.[9] The following description is schematic only: the key requirement is that the particles concerned should be able to travel from one place to another by two different routes. Electrons (or some other particles) are emitted by a source (on the left of figure 7.3) and pass through one of two apertures, after which some of them hit a detector (on the right in figure 7.3). This is attached to a counter, which keeps a record of how many electrons have hit the detector. Let us suppose that if only the upper aperture is open then 7 electrons hit the detector every second (on average) while if only the lower aperture is open the number is 9. The question is how many electrons hit the detector every second if *both* apertures are open together.

From a classical point of view the answer is 16, just the sum of the two previous numbers. If one is devious enough, one might argue that the number could be slightly higher. It is logically possible that an electron might go through the

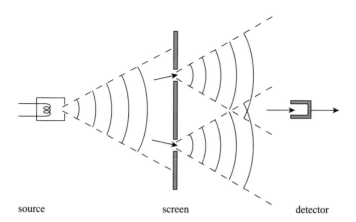

source   screen   detector

Fig. 7.3 The Double Slit Experiment

lower aperture, then back through the upper one and through the lower aperture a second time before hitting the detector; other more complicated possibilities could also be taken into account. The true, experimentally confirmed, answer is quite different. If the detector is in one of several calculable positions, no electrons enter it at all! By opening a second aperture a previously possible event becomes impossible.

This is so puzzling that it suggests trying to watch individual electrons as they pass through the apertures. Unfortunately this only serves to make the paradox deeper. If one makes any change in the experiment which enables one to determine which of the two apertures individual electrons pass through, then the number entering the detector changes to 16 per second as expected. Apparently each electron 'prefers' to travel through *both* slits simultaneously, but, if one tries to watch this happening, it stops doing so and behaves in a more normal fashion.

It has been suggested that the paradox is a group phenomenon: electrons are passing through the slits in a stream and those which go through one slit may be interacting with those which go through the other before any of them enter the detector. Unfortunately experiments show that this is not the explanation. If one reduces the flow of electrons steadily, until eventually there is surely only one in the apparatus at any time, the phenomenon is unaffected.

This beautiful but paradoxical experiment confirms once again that quantum particles are fundamentally different from their classical counterparts. However difficult it may be to imagine what is really happening, there is no doubt about the existence of the phenomenon. Nor is there any doubt that the use of the mathematics of quantum theory provides *quantitatively correct* predictions of the observations.

Very recently a Viennese group of physicists have observed the same effect for fullerene molecules composed of sixty or seventy carbon atoms.[10] Figure 7.4 is a picture of such a molecule. The molecule has a very definite structure, rather like that of a football, the corners being atoms and the edges representing chemical bonds. Some of the rings of carbon atoms in it consist of five atoms, while others contain six. It beggars belief that such an object might dissolve into two probability waves which later rematerialize as the original object, but that is the only way we have of explaining what is observed in words.

Quantum Probability

The 1930s saw not only Kolmogorov's formalization of probability theory but also von Neumann and Birkhoff's theory of quantum logic. This was an alternative probability calculus which described the newly discovered quantum theory. The history of this subject is rather unfortunate. The name quantum logic gives the impression that what was needed to come to terms with quantum theory was a new kind of logic. Over the following decades this idea was explicitly accepted by many researchers, many of whom knew less about the physics involved than

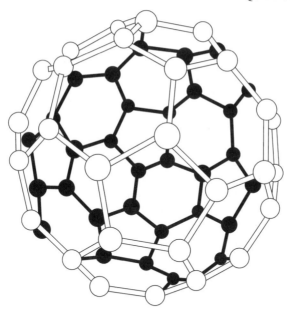

Fig. 7.4 The Fullerene Molecule C_{60}

was desirable. It now seems much more appropriate to view the revolution of quantum theory as being more to do with probability theory than logic. Segal's 1950 description of quantum theory in algebraic terms made clear the precise technical sense in which quantum theory deviates from Kolmogorov's probability theory. He wrote down a general algebraic formalism which included both classical probability theory and quantum theory as special cases. Kolmogorov's theory arose precisely when the algebra concerned had a very particular structure, but in quantum theory this was not the case and no such description was possible. In spite of the failure of Kolmogorov's axioms for quantum theory, the latter is clearly a probabilistic subject. Every introductory text tells one about wave functions, how they evolve in time, and how to extract probabilities from the wave functions. The process may seem extremely strange and counter-intuitive, but it gives correct predictions so reliably that it cannot be dismissed.

There was enormous resistance among classically trained probabilists to the idea that Kolmogorov's axioms were only one possible model of probability theory, and that it might not be applicable in some physical circumstances. The more common view was that his axioms *encapsulated what was meant by probability* so definitively that any situation in which they did not apply was by definition not probabilistic. Only in the last quarter of the twentieth century did this abhorrence for the paradoxes of quantum theory and its apparent denial of an observer-independent reality start to disappear. This was a revealing case in which many physicists were unwilling over a long period to distinguish between their mental models and the reality which they were supposed to describe.

Following the discovery of the many paradoxes of quantum theory, some physicists sought a classical account of the subject, in order to explain the probabilistic aspects of quantum theory in terms of Kolmogorov's standard account of probability theory. In this hoped-for reconciliation quantum particles would have straightforward positions and velocities at any moment, together with certain other parameters describing their internal structure. There should also be laws of motion which would describe how these evolve, and these laws would only involve influences on the particles from their immediate environments. More and more evidence accumulated that this was not possible, culminating in the discovery of Bell's inequalities, and the experiments of Aspect and others. Many strange explanations of these phenomena have been proposed, including the idea of influences which propagate faster than light but cannot carry any information. In the view of many scientists the cures of the supposed illness of quantum theory are no better than the malady. The current orthodoxy is that there cannot be any classical picture underlying quantum theory. It is, of course possible, some would say likely, that an entirely different way of looking at these phenomena will be discovered. This is likely to involve moving to ideas even stranger than those of quantum theory: the reinstatement of classical physics is an extremely unlikely scenario.

To summarize, Kolmogorov's axioms are no more a final definition of what mathematicians must mean by probability theory, than Euclid's axioms are a definition of what we must mean by geometry. The arrival of quantum theory destroyed Laplacian determinism even more thoroughly than did the discovery of chaos. It tells us that even if we were able to set up two experiments in *exactly* the same way their outcomes may well be quite different. Probabilities are embedded in the nature of the physical laws of the universe, and cannot be explained in terms of our ignorance of the necessary facts, *even in principle*.

Quantum Particles

In this section I will try to convey some impression of how a quantum particle is described mathematically. It is customary in this field to talk about the momentum of particles, but I will use the term velocity, which is more familiar. Before starting I must make an important point. Quantum mechanics is a mathematical model of reality. When we refer to a quantum particle, we will mean a particle as described by quantum theory. We do not intend to imply that quantum theory is *correct* by using this language, even though the mathematical model has survived all tests so far. Similarly when we distinguish between quantum and classical particles, we do not intend to imply that both types exist, and that they have different properties! Rather we are distinguishing between the predictions of quantum and classical *models* of a situation. Physicists frequently use language whose obvious interpretation is not the correct one. Thus when they say that they have proved that some new theory is true, they mean that the relevant mathematical model provides a much simpler or better match with reality than any previous ones.[11]

This is an important point, and is discussed further on page 265. Models are not required to represent reality perfectly, and simple models may often be used in preference to better but more complicated ones. Many physicists would say that we cannot ever know about reality itself, and have to content ourselves with constructing models of it which are simple enough for us to be able to understand. Others are more optimistic about our eventual ability to understand nature. But it cannot be denied that almost all of current fundamental physics consists of the production and testing of mathematical models.

In Newtonian mechanics a particle is considered to have an exact position in space at any instant. It also has other qualities, such as its velocity, mass and electric charge, which are all attached to the particle in some sense. On the other hand in quantum theory a particle is not located at a single point but has a shadowy presence throughout a small region. In the simplest possible case it has a phase and an amplitude at *every* point of space. The phase is an angle between 0° and 360° while the amplitude is a positive real number. The probability (density) that the particle is at some point is the square of its amplitude at that point, but the phase has no classical analogue. The total description of how the phase and amplitude of the particle vary from point to point in space is called its wave function or state.

If we forget about the issue of phase, it is easy to visualize a quantum particle in wave terms. A wave on the sea has an amplitude which varies from point to point, and decreases as one moves away from the centre of the wave. A water wave is not located at any particular place, but is concentrated around the region where its amplitude is greatest. One difference from quantum particles is that the total size of a water wave can be small or large. On the other hand if a quantum wave describes one particle and one sums (or rather integrates) the probabilities of the presence of the particle over all space points, one always gets a total of 1. A rather mysterious way of putting it is that a quantum particle is a probability wave.

The position of a quantum particle cannot be pinned down precisely from its phase and amplitude functions. If one follows the prescriptions of quantum measurement theory, one obtains a point in space, but this point is not the place where one is bound to find the particle if one looks. Rather it is the expected or average position near which the particle is likely to be. Everything in the subject is similar, reflecting the fact that the theory deals only with probabilities.

When two particles come together and interact their total local structure is not simply the sum of their separate local structures; the two local structures influence each other as time passes, leading to a composite structure for the pair of particles which cannot be disentangled. As the particles separate their local structures may continue to be entangled, so that the result of later measurements cannot be computed as if they were independent of each other. In some cases this entanglement remains even when the particles are far apart and cannot possibly interact with each other in any conventional sense. Entanglement is an intrinsically quantum phenomenon with no classical analogue.

Of course fundamental particles are too small to observe directly, and one may only infer their presence from events which can be seen at a macroscopic level. These could be bubbles in a cloud chamber, caused by the condensation of gas around an ionized atom, or the observation of a photon emitted by an atom as it changes from one energy level to another. The interaction which eventually leads to the observation involves energies comparable to those which the particle has in the first place, so the subsequent motion of the particle is substantially changed by the process. This is the famous uncertainty principle of Heisenberg: observations disturb the particle measured in a way which has a precise quantitative formulation.

This formulation of the uncertainty principle is not particularly disturbing— the same would apply to a very small classical particle. However, there is more. The position and velocity of a quantum particle are not well-defined even in situations where it is isolated and not subject to interactions or measurements. The particle is not localized at a single point but is spread throughout a small region. If the position distribution of a particular particle is very highly concentrated around a single point, then it turns out that its velocity distribution must be very spread out. There is a precise mathematical sense in which the position and velocity cannot both be accurately specified. This is the more fundamental uncertainty principle, which has nothing to do with observations. It forces one to recognize that quantum and classical particles are fundamentally different.

We make no conjecture about what fundamental particles themselves are really like, but content ourselves with knowing how accurate various mathematical models are in different circumstances. The quantum model is the best we currently have, but it is technically harder to use it than one would like.

The Three Aspects of Quantum Theory

The application of quantum theory in a typical laboratory setting has three stages. One first has to set up the apparatus in such a way that one knows the initial state of the particles concerned. This is called state preparation. The second stage involves writing down and solving the mathematical equations which describe how the particles evolve in time. The final problem is to calculate the results of the experiment from the wave functions, often called quantum measurement. Each of these steps is carried out using a standard collection of recipes.

Many of the deepest and most interesting results in quantum theory make no reference to state preparation or measurement, but concern only analytic properties of the mathematical model. We will refer to the study of these equations as quantum operator theory. This way of presenting the subject was the responsibility of Paul Dirac, but mathematicians would also want to mention the contribution of John von Neumann. Among the achievements of quantum operator theory are the calculation of the energy levels of atoms and molecules, which lies at the core of computational quantum chemistry.[12] Such results in quantum operator theory have led to great confidence in the relevance and predictive success of quantum theory. There is a high degree of agreement about what constitute correct procedures within it, because it is essentially a branch of mathematics, concentrating on a particular type of physically motivated problem. At present we have no other plausible candidate for explaining the many phenomena which quantum theory handles so well.

Quantum operator theory is particularly important when one has no control over the state of the system being investigated. This happens when one is applying quantum theory to a phenomenon originating outside a laboratory, for example natural radioactive decay or cosmic rays. Similarly, the computational study of complex biological molecules makes no reference to state preparation or measurement, and uses only the mathematical machinery of quantum operator theory. This type of approach to the design of new drugs has some claim to be the most practically important aspect of the subject.

The other aspect of quantum theory, relating to state preparation and measurements, is highly controversial. There are fundamental disagreements about the proper philosophical *interpretation* of the formalism, highlighted by a number of thought experiments, even when the technical *application* of the formalism is well understood. The use of Bayesian ideas in quantum theory is still controversial. By and large physicists are interested in the underlying laws, not the current degree of ignorance of experimenters. Nevertheless the former can often only be achieved via the latter. The moral is that the less an experiment invokes measurement theory the better, as far as physics is concerned.

Quantum Modelling

When one examines a typical text-book on the subject, one finds that quantum theory is like a tool-box with a manual. The manual contains general advice about the relevant part of mathematics and how systems evolve in time. The tool-box contains a collection of particular models which have been found useful in various contexts, together with a variety of procedures for extracting predictions from those models. It also provides tools for bolting together simple models in order to handle complex problems involving many particles and/or fields. Using the tool-box makes considerable demands on the quantum mechanic's skill and experience.

Let us examine how this works for the double slit experiment. A simple model of an electron passing through a double slit involves a wave function which has an amplitude and phase at each point of space. This neglects the internal structure of the electron. A better model would involve allowing the electron to have its two internal degrees of freedom (an electron is a spin 1/2 particle). Rather than considering individual electrons, one might consider the beam as a whole, in which case one must take account of their Fermi statistics and charges. It may be thought necessary to include some model of the walls of the slit, since the electron has some chance of being absorbed by the wall rather than passing through the slits. We should have some quantum mechanical model of the process of collision of the electrons with the eventual detector. Perhaps we should include the quantized electromagnetic field in the model, because it may have an effect upon the results.

Experimental physicists are well accustomed to problems of this type. They do a rough calculation to see whether the extra complication of including the above elaborations of the basic model makes an important difference. If it does, then the elaboration is included or the experimental setup is refined. The various effects referred to above are considered separately, but the most sophisticated model never gets close to consideration. Most likely the final experiment is described by using a series of separate quantum mechanical models, one for each part of the experiment.

The above process is absolutely standard, but bears little resemblance to ideas about there being an objective wave function in some 'correct' Hilbert space. What we actually have is a series of choices by the scientist, which are tested against experience with similar problems in the past. These choices involve the preparation and detection processes just as much as the dynamics. Scientists learn a set of procedures for constructing partial models, and these include rules about how to draw boundaries around the part of reality to be quantized. Scientists play a key part in the theory by making decisions about which model to use, based upon their knowledge of the subject and experience. There is no canonical choice of model, which must be simple enough for real calculations to be possible.

Nancy Cartwright discussed the status of model building in several areas of science in *The Dappled World*. In Chapter 8 she gave a careful description of the so-called BCS model of superconductivity, which won Bardeen, Cooper, and Schrieffer the Nobel Prize in 1972. She pointed out that the BCS model was not constructed from the fundamental Coulomb interactions between the electrons and nuclei involved. Instead the authors wrote down phenomenological equations which incorporated a variety of relevant effects. She concluded as follows:

> *We are used to thinking of the domain of a theory in terms of a set of objects and a set of properties on those objects that the theory governs, wheresoever those objects and properties are deemed to appear. I have argued instead that the domain is determined by the set of stock models for which the theory provides principled mathematical descriptions. We may have all the confidence in the world in the predictions of our theory about situations to which our models clearly apply—like the carefully controlled laboratory experiments which we build to fit our models as closely as possible. But that says nothing one way or the other about how much of the world our stock models can represent.*

The first part of this quotation is a fair description of the way in which quantum mechanics is applied. Her closing sentence might seem mere caution, but she makes it clear elsewhere that she is strongly opposed to the fundamentalist doctrine that laws discovered in the highly constrained setting of a laboratory must have universal significance. This is a provocative claim, and was strongly criticized by Philip Anderson in his review of her book.[13] I will explain in Chapter 10 why I do not endorse it.

Measuring Atomic Energy Levels

Let us consider the measurement problem for an atom which is in an excited state. An atom[14] has a discrete set of energy levels and may jump down from a higher one to a lower one, emitting a photon. In the most elementary description the jump is sudden and irreversible, unless another photon of the right energy arrives to push the atom back up to the higher energy level.

Reduction of the wave packet is often claimed to occur when one makes an observation to determine if the atom is in some particular energy level. One commonly says that asking this question forces the atom to change its state suddenly either to one for which the answer is yes or to one for which the answer is no. While the evolution of wave functions is generally continuous and reversible in time, performing such measurements is said to be discontinuous and irreversible. The measurement is said to cause a collapse of the wave function.

The above description of measurements is far too simple-minded. An atom cannot be observed directly. What one can do is direct a photon into collision with it, and observe its scattering or absorption. One can also observe the spontaneous emission of a photon. The true observation takes place at a very remote location (on the distance scale appropriate to atoms). There is a model of the process which includes both the atom and the quantized electromagnetic field, which carries photons away from the atom. Computing with this model is much more complicated than with the simple model previously described. It also gives a quite different understanding of what is happening. One starts with the atom in an excited state and no photons present. The wave function of the combined system evolves continuously and reversibly in such a way that the atom moves smoothly from the higher level towards the lower one and a photon emerges continuously from the neighbourhood of the atom. As time passes the probability that a photon has been emitted increases continuously towards 1 and the probability that the atom is in its higher energy level decreases towards 0.

Nothing above involves any act of measurement. Once the photon is far enough away from the atom it may interact with a photon counter. If one knows that the photon has interacted with the photon counter then it becomes appropriate to use a different wave function to describe it. This has already arisen in the classical context, where we agreed (I hope) that probabilities may describe one's information about a situation rather than the situation itself. So with quantum theory. The wave function describes our best knowledge of the system: if our knowledge changes then we should use a different wave function.

An objection to this description of events is that it does not discuss the mechanism by which the information about the atom changes from being potential quantum mechanical information into objective fact. The answer is that quantum theory is a set of mathematical models of various degrees of sophistication which allow one to make predictions. Particles are not the same as the probability wave functions which we assign to them in quantum theory, and the question presupposes this identification. There is a real world, but we should not imagine that our best current model is more than that.

The EPR Paradox

Physicists frequently seek situations in which their theories make paradoxical predictions. In cases which can be realized experimentally, one of two things may happen. The prediction may be wrong, in which case their theory collapses, or at least needs to be modified. Alternatively the prediction is borne out, and they need to think about what is wrong with their intuition. Once they have come to terms with the unexpected phenomenon, the paradox ceases to be one. Unfortunately it sometimes continues to be called a paradox simply out of tradition.

The EPR paradox is a beautiful example of this. It was devised by Einstein, Podolski, and Rosen in 1935 to prove that there must be something fundamentally wrong with quantum theory. Unfortunately for them, it did nothing of the kind. The behaviour predicted by the model was precisely what was observed to happen in experiments, the most definitive of which were carried out by Aspect in 1981. This fact indicates that our naive mental images of quantum particles fail to correspond to reality in a rather fundamental manner. As time has passed physicists have come to terms with the phenomenon, although there are still many who feel uneasy about it. So Penrose has devoted several lengthy discussions to its supposedly paradoxical status, while Gell-Mann, has robustly dismissed these.[15]

The paradox involves a property of electrons called their spin. To a first approximation one can think of this as a little arrow attached to each electron, whose direction is the axis around which the electron is spinning. The paradox relates to the fact that in some respects an electron behaves as if the spin axis really exists, but appropriately designed experiments demonstrate that it cannot. Let us consider Bohm's reformulation of the paradox. We suppose that some atomic event leads to the emission of two electrons, which travel away from the atom in opposite directions (a similar discussion applies to photons). We assume that the combined spins of the two electrons have the simplest possible configuration. This is a pure, rotationally invariant, quantum state, often written in the form

$$\psi = \uparrow\downarrow - \downarrow\uparrow$$

Many more fanciful forms of this equation have been written, including one of Penrose in which the two types of arrow are replaced by little pictures of dead and live cats!

The natural interpretation of the state ψ is that either the spin axis of the first electron points in the direction ↑ and the second in the direction ↓, or vice versa; the two cases are equally probable. One might say that the spin axes of the two electrons point in the same direction, but one electron spins clockwise and the other anti-clockwise, so the spins cancel out. Which electron spins clockwise and which anti-clockwise is a matter of chance.

Unfortunately the mathematics of spin one-half particles allows one to write the *same* state in the form

$$\psi = \overset{\rightarrow}{\leftarrow} - \overset{\leftarrow}{\rightarrow}$$

In this form the natural interpretation is that either the first electron is in the state ← and the second is in the state →, or vice versa; once again the two cases are equally probable. Now up-down alternatives for the spin axis are quite different classically from left-right alternatives. There is no classical way of reconciling them, but quantum mechanically they co-exist.

The presence of the minus sign in the above equations provides another indication of the enigmas of quantum theory. There is no classical or probabilistic meaning to the idea of *subtracting* one possible configuration, ↓ ↑, from another, ↑ ↓. In quantum theory not only is this possible, but the result is experimentally distinguishable from the result of *adding* them together. Even more confusingly one *cannot* distinguish experimentally between ↑ ↓ − ↓ ↑ and ↓ ↑ − ↑ ↓. Because the two configurations are combined in this manner, the electron spins are said to be entangled. This concept has no classical analogue, and implies that one cannot regard the two electrons as independent particles, even when they are widely separated.

The EPR paradox is often described in terms of the results of measurements performed on one or both of the electrons. We will bypass these, since what we have already seen implies that the results *must* be paradoxical. If the state itself has non-classical properties, then there is no way in which measurements can eliminate this fact. In an earlier version of this book I had given a more detailed technical discussion of the EPR paradox, but was persuaded that this was pointless. The main protagonists actually agree about the mathematics involved. They also agree about the physics. Their problems relate to the philosophical status of the theory, not to the theory itself. Those in the first camp feel that there must be an objective state of affairs, and continue to seek either a new interpretation of quantum theory or a modification of the theory which will permit this. Those in the second camp regard discussions about the underlying reality as lacking meaning. They are completely satisfied with a mathematical formalism which provides correct predictions.

My own position is closer to the second camp. I do not consider that the mathematical description provides a full understanding, but accept that it may be the best which our type of minds are capable of. In the light of the known paradoxes, it is surprising that so many people hope that one day we will find a new formulation of quantum theory which will also correspond to our native intuition. I do not believe that God, if he exists, is a mathematician. Surely he would be vastly amused at our idea that such a tool might one day encompass the vastness of his creation. Did prehistoric people analogously believe that their gods had created the world using extremely delicate and sophisticated stone axes?

Reflections

In order to understand better the tension between the quantum and classical pictures of the world, one needs to talk to chemists rather than physicists. The former have a more thorough understanding of the issues because they have to deal with them daily! In the standard ball and stick models of molecules, such as that of the fullerene molecule drawn earlier, the balls indicate the positions of atoms, while the sticks represent bonds between atoms. Quantum theory tells us that the bonds are in fact *very* crude representations of forces due to smeared out electron wave functions. This crude classical model of molecules works astonishingly well, and is still widely used today. However, it fails to explain many issues, such as the properties of non-rigid molecules.

In this section we will discuss chiral molecules. These are molecules which appear in two forms which are simple reflections of each other, in the same way as left and right hands are. Many well-known compounds, such as glucose and vitamin C, have two such forms, which have very different biological activity. In the case of thalidomide, one of the two forms acts as an effective sedative, but the other can produce major embryonic malformations. Unfortunately, when it was used as a drug in the 1960s, the two forms were mixed together, with disastrous effects.

If one looks at the structure of molecules from a (naive) physicist's point of view one would say that chiral molecules should not exist. According to early views of quantum theory molecules are always in states with mathematically sharp energies and occasionally make sudden transitions between these by absorbing or emitting a photon. Because the laws of quantum theory are invariant under reflections the sharp energy states must also be so. Therefore they cannot be chiral.[16]

There have been several attempts to resolve this paradox. It has been suggested that some future development of quantum theory will incorporate nonlinear effects. These would have to be extremely small, since quantum theory is so well verified, but they might allow the existence of stable chiral molecules. It has also been proposed that a known weak force which is not invariant under mirror symmetries may be important in this context. In my opinion both speculations are the wrong direction to seek the explanation: the effects are too weak to have any significant influence, except possibly in interstellar space or extremely rarified gases. In all normal situations molecules are subject to constant collisions with their neighbours, and this will overwhelm any much weaker effects, even if they exist.

The actual flaw in the argument is the prejudice that atoms are always in sharp energy states. This is a convenient fiction which makes calculations much simpler. But it is well known to chemists that the appropriate state of a molecule depends upon how it was prepared. The manufacture of chiral molecules depends upon using a process which selects the left- (or right-) handed form preferentially. The same applies to biological molecules, many of which exist in only one of the two possible forms. It is a mistake to look for some fundamental explanation of this. All current organisms on Earth probably evolved from a single ancestor, and it is entirely plausible that the handedness of its constituent molecules was simply a matter of chance. From that point onwards evolution saw to it that only copies of those molecules, necessarily with the same handedness, were manufactured. History does matter!

The fiction that molecules are always in sharp energy states is also responsible for the idea that there must be a sudden change in the state of a molecule when it is observed to 'jump' from one energy level to another. In fact when one does the full calculation of the interaction between a molecule and its environment all changes are continuous. This is not a philosophical point: chemists could not understand the dynamics of chemical reactions without carrying out these more complete calculations, which involve the consideration of states which do not have sharp energies.

Schrödinger's Cat

More heat has been expended on the problems of quantum measurement than on any other aspect of the subject. It has been claimed that in the process of measurement a quantum particle undergoes a sudden collapse: its state suddenly changes from its pre-measurement value to its post-measurement value. In the last section I explained why this is wrong by referring to standard ideas taught to all chemistry students. Here I discuss the famous story of Schrödinger's cat.

In this (thought) experiment a cat is sealed inside a room with some apparatus. The decay of a radioactive atom in the room is detected by a Geiger counter and triggers the death of the cat. The fate of the cat and of the atom cannot be disentangled, and the radioactive decay can only be explained in quantum mechanical terms. It is then argued that at a deep enough level the cat's fate must be also described using quantum theory. Indeed while the room remains sealed the cat is supposed to be suspended in a quantum state, partially dead and partially alive. This state is beyond intuitive understanding, even though its mathematical description is simple and unambiguous. One might say that the cat is in limbo, neither properly dead nor alive. When someone enters the room and observes the cat this forces a collapse of the wave packet. The cat suddenly becomes actually dead or alive.

I do not suppose that many people actually believe that cats can actually exist in limbo, returning to the real world when observed. If one did then one would be led into a series of ever more implausible questions. Who is

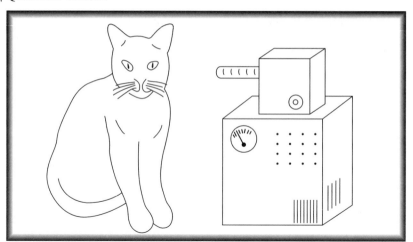

Fig. 7.5 Schrödinger's Cat

competent to collapse a wave packet and who is merely a part of the wave function of other more significant observers? Does Schrödinger's cat itself have enough consciousness to collapse wave packets? Is quantum theory really about dividing the animal kingdom or even the human race into those who are effective observers and those who are not? Does the collapse take place when the relevant photons hit our eyes or only when the signal reaches our consciousness? How can one possibly answer these questions on a scientific basis?

Several different resolutions of the paradox have been proposed. Most of them have some merits. The following three solutions are very much a personal selection. They are closely related and may be complementary ways of describing the same truth.

The first solution is to deny that cats can be quantized. We have already pointed out on page 66 that cats do not have well-defined boundaries. Exactly what atoms are part of an individual cat is not objectively decidable: the number keeps changing as the cat breathes in and out or digests its last meal. This being so it is impossible to assign a quantum state to a cat. The situation is very different from that for the fullerene molecule discussed previously, which has an absolutely precise number of atoms locked into a particular geometric configuration. If one replaces the cat by an object simple enough that one can plausibly assign a quantum state to it, then the methods of quantum theory do indeed become applicable. If one represents the cat by a simple two-dimensional dead-alive quantum system, then one needs to remember that this is an outrageous abstraction of the cat itself.

A second approach. Reduction of the wave packet has nothing to do with consciousness. The supposed reduction of the wave packet is not an objective effect but the result of giving an approximate description within an highly simplified model. In this simplified model the effects of the environment are ignored apart from the inclusion of a wave-packet reduction formula. There is no point at which one can make a principled division between the object studied and the environment, but one has to make it at a point at which the quantum mechanical model can *actually* be solved. This point is not sharply defined, but it occurs long before one has got near the consciousness of any observer or of the cat. I quote Omnes concerning the reality of the reduction of the wave packet:

> *However there is no physical effect that might be called a reduction effect*
> *… No formula resulting from a mathematical analysis is supposed to have a*
> *physical content, and wave-packet reduction is only a formula expressing the*
> *result of a calculation in logic.*[17]

A final attempt. The quantum state is never attached to the particle or system which it describes, but is rather our best (current) way of encapsulating the information we have about it. If two people have very different knowledge about an entangled quantum system, possibly because they are physically separated, then they are right to use different states to describe it. In the words of the physicist Roger Newton 'the befuddlement arises from the mistaken notion that a quantum-theoretical state, as described in the ideal case by a wave function, is a direct description of reality'.[18] Cats should not be confused with the mathematical formulae which we use to predict the results of future observations of them. Once the cat has been observed it becomes appropriate to use a different mathematical state to describe it. As with the EPR paradox, the state which we use to describe something depends upon the information we have.

Let me sum up. It might seem trite to say that quantum theory is a mathematical theory, which should not be confused with the reality it claims to describe. However, many of the paradoxes of quantum theory only arise because of people's failure to make precisely this distinction. The quantum state is a mathematical construct quite distinct from the physical particle, about whose nature we know little. Quantum theory makes extremely accurate predictions in some simple situations and gives a good understanding in many other more complicated ones. The equations are not always soluble in any practical sense, and we have no proof that they apply in extremely complicated and dynamically unstable situations. We do not understand why the quantum mechanical equations work. Perhaps the truth is just that we have kept on seeking equations which describe various natural phenomena, and it would have been surprising if we had not had considerable success, at least for some of those phenomena, after several centuries of continuous effort. In another few centuries we should expect that the equations then used will be even more effective. The universe must not be identified with some set of equations which humans invented, and can only solve with a good degree of accuracy in very special circumstances. The phenomenon of quantum entanglement indicates that the universe cannot be decomposed into small isolated parts. This has the fundamental consequence that the way science has always operated is, finally, misconceived. What is astonishing is that we have got so far by studying small parts of the universe in isolation from the rest.

Notes and References

[1] Gillies 2000b

[2] Jaynes 1985

[3] Brush 1976b, p. 682

[4] It is conventional in mathematics to express probabilities as numbers between 0 and 1. To convert them to percentages multiply by 100.

[5] This is known as the epistemological interpretation of probability, and is frequently the most compelling in spite of its apparently paradoxical nature.

[6] No doubt most of my mail relating to this book will be from people who condemn my narrow-mindedness about these issues, but one has to live with such things.

[7] Brush 1976a, p. 292

[8] Einstein's explanation of the photoelectric effect in 1905 was cited when he was awarded the Nobel Prize for his contributions to physics in 1921.

[9] Dürr et al. 1998

[10] Brezger et al. 2002

[11] I should not claim to speak for all physicists: some do indeed think that their equations 'are obeyed' by reality, in spite of the fact that there is no consistent integration of quantum theory with general relativity.

[12] Other notable achievements are the spin-statistics theorem in the Wightman theory, the recent proof of asymptotic completeness for non-relativistic N-body scattering by Sigal and others, the Lieb–Thirring work on the stability of matter, and the ongoing study of the Anderson model via the spectral analysis of random Schrödinger operators.

[13] Anderson 2001

[14] I use this word as an abbreviation for the nucleus together with its cloud of electrons. The energy levels of the latter are clearly the key issue here.

[15] Penrose 1989, Penrose 1994, Gell-Mann 1994

[16] I use the term states with sharp energies in place of the technical term eigenstates. Technically speaking every eigenstate of a linear quantum mechanical Hamiltonian must be even or odd under space inversion symmetries. Chiral states are not.

[17] Omnes 1992, p. 364

[18] Newton 1997, p. 186

8
Is Evolution a Theory?

Introduction

In Chapter 6 we explained how one of the most impressive edifices in physics, Newton's laws of motion, was superseded early in the twentieth century by other theories (quantum mechanics and general relativity) relying upon completely different mathematical foundations. The possibility of such a dramatic upheaval appears to support cultural relativism, the idea that scientific theories are culturally determined, like any other belief system. This stands in strong contrast to the belief of scientists that they are progressively discovering objective facts about the natural world. The distinction between the subjective and the objective is indeed hard to define, and is nowhere more controversial than in the subject of this chapter, evolutionary theory. It is clear that one cannot have a balanced view of the philosophy of science if one's only input is the way mathematicians and physicists view their subjects. Unfortunately the philosophy of science is dominated by physics, partly because most philosophers lived before the main era of development of the biological sciences. Even in the twentieth century many writers seem to support the view that the depth of a science is determined by the extent to which it depends on mathematics. Here we should distinguish between those sciences which use mathematics but whose theories can be phrased in common-sense terms, and others, such as quantum theory, which depend entirely on mathematics for their formulation. Biology and geology are of the former type, but this does not prevent them being amongst the most interesting scientific subjects which the human race has investigated.

Since 1980 a large number of excellent books on biology and the theory of evolution have been published. Most of the authors have no interest in or respect for creationism, and concentrate on communicating the fascination of their own field. Our focus will be somewhat different. After discussing briefly how scientists came to their present beliefs about the origin of species, we will concentrate on establishing which parts of evolutionary theory are established fact and which parts are still hypothetical. Popper has told us that no scientific knowledge is ever final, but many philosophers and scientists do not share his scepticism. His views will be discussed in greater detail in the final chapter.

Before proceeding, I must emphasize that biology does not have the tight logical structure which physical scientists wrongly regard as the hallmark of proper science. Although in one sense shallower than physics, it is far broader. Its theories gain conviction from the huge variety of supporting evidence, not because of the existence of critical experiments. This chapter follows the same pattern as others: while the ostensible subject matter is the theory of evolution, the actual focus is on how one makes decisions about the objectivity of scientific knowledge.

The Public Perception

The publication of Charles Darwin's *The Origin of Species* in 1859 provoked an intense debate about its scientific merits and implications for the Christian faith. The eventual outcome in the United Kingdom was an acknowledgement that literal creationism was not a tenable doctrine, and that human beings have indeed evolved from apes. The Church of England has no problems with this, and senior members of the Church often speak in support of evolution.

The situation in the United States is entirely different. Opposition to evolution has grown within the fundamentalist Protestant community there, particularly during the second half of the twentieth century. A series of Gallup polls held between 1982 and 2001 all reveal results very similar to that of 1999. Americans favour teaching creationism in the public schools along with evolution by a margin of 68% to 29%, and 40% even favour *replacing* evolution by creationism. About 40% of Americans believe that human beings have developed over millions of years from less advanced forms of life, but God has guided the process; 9% believe that human beings have developed over millions of years from less advanced forms of life, and God had no part in this process; finally 47% believe that God created human beings pretty much in their present form at some time within the last 10,000 years or so. I will refer to this last group as hard-line creationists.

The position of the Catholic Church is clear if, as usual, very cautious. In October 1996 Pope John Paul II made the following statement:

> Today ... new knowledge has led to the recognition of the theory of evolution as more than a hypothesis. It is indeed remarkable that the theory has been progressively accepted by researchers, following a series of discoveries in various fields of knowledge. The convergence, neither sought nor fabricated, of the results of work that was conducted independently is in itself a significant argument in favour of this theory.

He did not, however, support Darwin explicitly, declaring, quite reasonably, that 'to tell the truth, rather than *the* theory of evolution, we should speak of *several* theories of evolution'.

Many scientists either ignore or ridicule the creationists, but this has not made them disappear, and we will examine the strength of their arguments in this chapter. This chapter will explain why some aspects of evolutionary theory have now passed into the corpus of settled knowledge about the world. I will follow the standard practice of using the term 'evolution' when referring to the appearance and extinction of species over a time scale of many millions of years. The phrase 'theory of evolution' in later sections refers to the more problematical issue of whether the mechanism driving evolution was what Darwin proposed. The distinction between these two issues will be of key importance. The next two sections describe a small fraction of the evidence that the Earth is indeed several billion years old. The only way of avoiding this conclusion is to suppose that vast quantities of false evidence have been deliberately planted by a being with supernatural powers.

The Geological Record

Speculations about the nature and origins of fossils started over two thousand years ago, and included a particularly perceptive analysis by Leonardo da Vinci. However, the serious investigation of fossils started with Georges Cuvier in Paris at the end of the eighteenth century. He was the first to think of comparing the anatomy of fossilized bones systematically, and to record the rock formations from which they were taken. The conclusion was inevitable. Many species of large animals had become extinct, and fossils discovered nearer the surface were always closer in form to existing species. In 1812 Cuvier published four volumes explaining the significance of a vast number of careful observations, from which it became clear that the Earth could not be just a few thousand years old.

From this point the science of stratigraphy developed rapidly. In England William Smith used the steady progression of forms of life in the fossil record to classify the rock layers in his book *Strata Identified by Organized Fossils* published in 1816. Mary Anning devoted her life to the collection of magnificent and often complete fossils of plesiosaurs, ichthyosaurs, and other prehistoric reptiles. Some of her specimens, which are up to eight metres in length, may be seen in the Natural History Museum in London. The fossil in figure 8.1 is about two metres long and dates from about 200 million years ago. Of course nobody could have known their age in the nineteenth century, but it was obvious that they bore little resemblance to any living creatures. Richard Owen was responsible for the first attempt to build models of extinct animals based upon their fossils, at an exhibition in Crystal Palace Park in South London which opened in 1854. The models may still be seen there. So it was well known to the public, long before 1850, that animals quite unlike those we now know had once existed and perished.

In the first half of the nineteenth century, the accepted explanation for the above, promoted by such authorities as Cuvier and Richard Owen, was called catastrophism. According to this, animals retained their forms unchanged until they suddenly became extinct in some natural catastrophe, to be replaced by quite new forms in a separate and independent act of creation. The fixity of

Fig. 8.1 Plesiosaurus hawkinsii
© The Natural History Museum, London

species was supported by the fact that there had been no changes in known species throughout recorded history.

Two people did, however, argue that species might evolve from one form to another over a long enough period of time. The first was Erasmus Darwin, the grandfather of Charles Darwin, who published a two-volume work called *Zoonomia* on his theory of evolution in the last decade of the eighteenth century. Apparently ignorant of this, Jean-Baptiste Lamarck published a seven-volume theory of evolution between 1815–22. His ideas, which are discussed further on page 213, were considered obviously false, and were not followed up when he died shortly afterwards at the age of 78. The time was not yet ripe for such ideas to be taken seriously.

A lecture of Thomas Huxley given to the British Association in Norwich in 1868 described some of the evidence for the vast time scale needed for the evolution of life and rocks. He explained that there is a layer of white chalk, hundreds of metres thick in places, stretching from north-west Ireland to the Aral Sea via the Dover cliffs, central Europe, and Syria. When examined under a microscope it is found that the chalk is composed of the shells of tiny organisms called *Globigerinae* together with other organic granules called *coccoliths*. These are remnants of marine organisms, indicating that this whole area was once at the bottom of the sea. Such a thick layer must have taken an extremely long period of time to accumulate, only to be covered by yet other layers of rock. All this had to happen before this ocean floor started to rise ever so gradually above sea level. Whether the time involved should be measured in millions, tens of millions, or hundreds of millions of years is a technical question, but it is clear that such a process could not occur within a few thousand years.

The next item in my selection is the theory of continental drift. Although related ideas had been proposed on several occasions earlier, the first detailed theory of continental drift was put forward by Alfred Wegener in 1912 and then more fully in *The Origin of Continents and Oceans* published in German in 1915. He provided evidence that our present continents had been formed by the break-up of a single supercontinent, Pangaea. His ideas went far further than just accounting for the similarity of the Western coastline of Africa and the Eastern coastline of South America. There is a striking relationship between the geological formations on either side of the South Atlantic Ocean—one for which we now have vastly more evidence than he did. In addition the fossil records on the two sides of the Atlantic up to the presumed era of separation have strong similarities. To give just one isolated example of Wegener himself, fossils of a Permian[1] reptile called the mesosaur may be found only in South Africa and Brazil. Since this was a freshwater species living in lakes and ponds, it is extremely implausible that it could have made its way across a divide of several thousand kilometres of sea.

The Origin of Continents and Oceans contains far more material than described here. Much of it is very convincing, but the chapter dealing with actual evidence for the movement of the continents is not: geodesy was simply not accurate enough in his day to demonstrate the very slow movements taking place. Nevertheless, there was abundant evidence in support of his theory, and it is interesting that Wegener's ideas did not acquire scientific respectability until long after his death. The reason is surely that there was no plausible mechanism by which continents could move physically over the Earth's surface. This being the case, people preferred to believe that the evidence was surprising but not ultimately significant.

In the 1960s the situation changed dramatically. Surveys of the bottom of the Atlantic identified a ridge in the mid-Atlantic (with similar ridges on the ocean bottoms elsewhere in the world). Hess proposed that basaltic magma was being pushed to the surface at the ridges and that the sea floor on either side of these ridges were moving away from it. Evidence in favour of this idea was not long in appearing. It included the following:

- Observations of the mid-Atlantic ridge using underwater cameras showed molten rock emerging all along it up to Iceland, which is one of the most active volcanic regions in the world.
- Magnetic field reversals are recorded in the rocks in roughly parallel lines on either side of the central divide; this has been explained on the basis that both sides are moving steadily away from each other as new material is pushed to the surface at the divide itself.
- The age of the ocean floor increases with its distance from the mid-ocean ridges, but is no more than two hundred million years anywhere—a tiny fraction of the age of the land masses.
- The continuing separation of the continents at the rate of a few centimetres per year on either side of the Atlantic has been measured using orbiting satellites.

- The rate of separation is consistent with the date for the separation of Pangaea obtained by dating methods based upon radioactive decay, namely that it occurred over the period between one and two hundred million years ago.

It would take a whole book to describe the historical process which took the new theory of 'plate tectonics' from the realm of an implausible story in 1960 to that of widely agreed fact by the early 1970s (and several more to describe the present evidence in support of the theory). Any such account would have to describe the important contributions of the research vessel Glomar Challenger, launched in 1967 to take cores from the ocean floor. Agreement was reached because new evidence from several independent fields of science supported the same overall picture, *and also* because people finally understood the mechanism behind the continental drift.

My next choice is much more visible. The rocks in the Grand Canyon in Arizona naturally divide into twelve layers, most of which are between 100 and 200 metres thick. Let us consider just one of these, the Bright Angel Shale, which is the tenth counting downwards. This is a greenish gray shale containing fossils of marine arthropods called trilobites. The presence of the fossils establishes that this layer is a sedimentary deposit. The vast extent of the layers, which are not confined just to the canyon, attests to a process which could not have occurred in a few thousand years. Moreover, all of the layers in this plateau must have been formed when the area was below sea level and long before their erosion by the Colorado river and the weather. Considering the many cubic kilometers which have been removed by erosion, the erosion process could not possibly have taken only the few years of a Biblical flood. It is believed by experts that the formation of the canyon actually took somewhere between one and six million years. The rocks themselves were laid down far earlier. Trilobites first appeared about 540 million years ago, and after dominating the seas for long periods, eventually became extinct by 250 million years ago.

Let us return to the remote past. The Earth is believed to have been formed 4.6 billion years ago, and the earliest known rocks on the Earth are about four billion years old. The first primitive lifeforms were well established by 3.5 billion years ago. There was an explosion of new species in the late Precambrian era about 570 million years ago, and over the next 300 million years many animals and plants were established on land. It seems that simple, single-celled organisms arose relatively quickly as soon as the conditions made it at all possible. The more remarkable (because the longer delayed) event was not the appearance of life but the evolution of multi-cellular organisms with complicated internal structures. The appearance of new species since that period has been anything but steady. There have been three major extinction events dating around 65, 200, and 250 million years ago. The most recent is the best known, and coincided with the disappearance of the dinosaurs. I will discuss instead the evidence for the middle one.

The Triassic period came to an abrupt end some time about 200 million years ago, ushering in the Jurassic era. The latter is named after the Jura mountains in Switzerland, where the 'Jurassic' rocks were first studied by Alexander von

Humboldt in 1799, but in fact the transition was a world-wide event. Its absolute date has been narrowed down to between 199 and 203 million years ago by measuring radioactive decay products. Many scientists have devoted their careers to trying to find out what precipitated the event, and there is now a project to take systematic rock cores over the world to resolve these questions. But already a great amount is known, with contributions from several different fields. The fossil record shows the disappearance of about a half of all species of land animals. Pollen and spore fossils and marine shelly organisms show similar abrupt changes.

It has recently been observed that there was a dramatic change in the density of stomata on the leaves of plants from the same period.[2] Comparisons with the stomata densities for the closest related present day species indicate an increase of carbon dioxide concentration from 600*ppm* to 2300*ppm* across the Jurassic-Triassic boundary. This could have raised the global temperature by 3 to 4 degrees centigrade. There was also a major reduction in leaf sizes during this period, explicable as selection to avoid lethal leaf temperatures.

The probable cause of these events has also been identified. There are three huge areas of volcanic rocks dating from almost exactly this time covering almost three million square miles in Brazil, West Africa, and North East America. The original lava flows would have released carbon dioxide into the atmosphere in enormous quantities. The three areas involved are widely separated now, but at the time they were parts of one area in a supercontinent called Pangaea. This area, called the CAMP, is bounded by the broken lines in figure 8.2, which also indicates the approximate position of the three continents 200 million years ago. The break-up of Pangaea eventually led to the present positions of the continents by the process of continental drift. It is not known whether there was also a precipitating event, such as the impact of a large comet, and much remains to be done.

The investigation of the causes of this extinction event has not yet got to the point at which it is beyond dispute. The interesting issue from our point of view is the methodology adopted for resolving the problem, namely the search for *several independent sources of information*. In this case the evidence comes from three different continents, and the rate of progress suggests that the outcome will be clear within ten years or so.

Dating Techniques

All of the examples in the last section attest to the great age of the Earth, but much of our more detailed knowledge depends upon accurate dating techniques. These are not straightforward, and require painstaking comparisons between independent approaches to this problem. An idea of how this process is carried out is given by the following example, which relates to the modern era.

During the last thousand years five supernovae are known to have occurred from the historical record, in 1006, 1054, 1181, 1572 (Tycho Brahe), and 1604 AD (Kepler). The 1054 supernova, found in Arabic and Chinese records, is associated with the now famous Crab nebula. (Many more have been catalogued

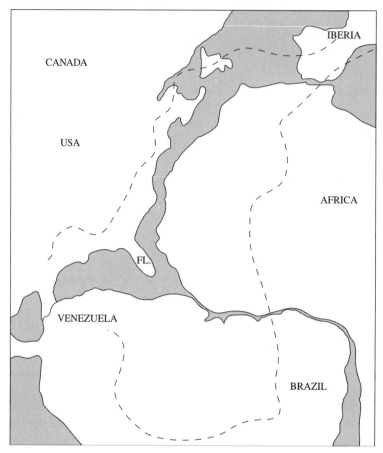

Fig. 8.2 The Central Atlantic Magmatic Province[3]
Adapted from Marzoli et al. 1999, to be found at http://jmchone.web.wesleyan.edu/

in recent times, but they were too faint to have been seen with the naked eye.) Each of these can be identified with objects which can now be seen only with the aid of powerful telescopes. Our knowledge of the physics of supernovae confirms that they exploded at the observed times.

In 1998 the X-ray satellite ROSAT reported the remnant of another supernova, which must have exploded some time early in the fourteenth century. Although no written record of it has yet been found, there is an entirely different source of evidence of the event. An analysis of the nitrate concentration of a hundred metre section of an Antarctic ice core has revealed four sharp peaks along its length. These corresponded to dates around 1180, 1320, 1570, and 1600 AD. The beautiful match between the three totally independent sources of information provides good evidence that each of them is reliable. There is even a mechanism for the increased concentration of nitrates in the ice. The explosion of a supernova bathes the Earth's atmosphere with ionizing radiation, and this would lead to the reaction of nitrogen with oxygen to produce nitric oxide and then nitrates by known chemical processes.

The dating of events thousands or even millions of years ago is a more technically complicated matter. The most precise, tree ring dating, gives exact dates for climatic events up to about ten thousand years ago. Ice core dating is reasonably precise and has the advantage that Antarctic ice cores have been extracted with ages of up to four hundred thousand years. The dating of rocks over even longer time-scales depends largely upon isotopic analysis, or radioactivity. The latter phenomenon was quite unknown during the nineteenth century, when most geological dating was essentially guesswork. One of the main difficulties was that sedimentary layers of rock can only be formed at places which are submerged under the sea. It follows that major sedimentary layers may be entirely missing in one part of a continent, because that area was above sea level for tens of millions of years.

Let me give a potted history of the discovery of radioactivity. In 1896 Becquerel discovered that certain uranium salts were able to affect a photographic emulsion by means of an unknown and invisible type of radiation. This spurred Marie and Pierre Curie to discover the radioactive elements polonium and radium, with the result that all three shared the Nobel Prize for Physics in 1903. Rutherford explained radioactivity as the result of the disintegration of unstable elements into lighter and more stable ones, with the associated emission of energetic particles, in *Radioactivity*, published in 1904. These events were the prelude to a transformation of our understanding of the nature of matter, and to the creation of several new industries; these relate to atom bombs, nuclear power stations, new medical diagnostic tools, and treatments for cancer. There are two related ways in which it has altered our understanding of the history of the Earth.

The nineteenth century physicist Thomson estimated the age of the Earth, assuming that it contained no internal source of heat and had gradually cooled down to its present state from a much hotter initial condition. Over a period of a few decades he refined his calculations to reach the conclusion in 1899 that the age of the Earth must be between 20 and 40 million years. This was regarded by geologists as being far too short to accommodate their observations, and neither side would give way. The discovery of radioactivity solved this problem. The Earth was not inert, and radioactive elements such as uranium were sufficiently common to provide an important source of heat within the Earth's crust. This allowed the Earth to be much older than had previously been thought by the physicists, and brought them into line with the geologists.

Radioactivity has also affected our understanding of the history of the Earth by providing values for the ages of rocks. Its reliability in this respect can only be understood in the context of its basis in physics, so we will start there. Matter is composed of about a hundred different elements, most of which come in two or more forms, called isotopes, with slightly different weights. The main form of carbon is called carbon 12, but there are two other isotopes, carbon 13 and carbon 14, both of which are radioactive. The rate at which an isotope decays into other elements is determined by its half-life: this is the length of time it takes a half of the atoms in a sample to decay.

Most elements occurring in nature are stable over periods of many millions of years. This is hardly surprising, since if they were not the atoms would already have decayed into something else. The different isotopes of an element usually occur in different proportions: for example uranium 235 makes up only 0.7% of all naturally occurring uranium, almost all of the rest being uranium 238. The dating of rocks depends upon the fact that certain isotopes of carbon, uranium, potassium, and other elements are slightly radioactive and decay at known rates to other elements. By comparing the amount of the decay products with the amount of the original element in a sample, the age of the sample can be calculated.

There are many reasons why the procedure just described might not yield correct results. Here are just a few:

- The rate of decay of radioactive substances might have been different in the past from what it is now.
- The original sample might have contained some of the decay products naturally, with the consequence that the present amount does not depend solely on the age of the sample.
- During the lifetime of the sample either the original element or the decay products might have diffused into or out of the sample.
- In the case of the decay of carbon 14 the source of this isotope is supposed to be the cosmic ray flux in the upper atmosphere and this might have varied during the 50,000 years for which this particular technique is useful.

The above possibilities are not ones which I have invented for the purpose of discrediting radioactive dating techniques. In fact they are only a fraction of the issues which scientists have themselves raised and found ways of resolving. Many ingenious methods have been used to test the reliability of the dating techniques. Of particular importance is cross-checking between independent methods of dating to see if they yield the same values. A recent monograph of Bradley[4] devotes 610 pages to the systematic comparison of dating methods for the Quaternary Period (the last 1.6 million years) alone!

One of the methods of testing the reliability of dating methods uses the decay of potassium 40 to argon 40. It depends upon the fact that both potassium and argon have other naturally occurring isotopes. One can compare the apparent age of the sample as calculated from each of the isotopes separately. If these are consistent it increases one's confidence that the age calculated is correct. If one chooses a crystal free of obvious defects this increases the likelihood that it was formed at a single time. If the concentration of the various isotopes throughout the crystal is constant, this provides further evidence of the same type. If, however, the concentrations of the isotopes vary between the surface layers of the crystal and its interior, this suggests later contamination and may force rejection of the sample. The physics of this and many other tests of the dating procedures have occupied scientists for many years, and as time has passed they have steadily improved the reliability of the tests which they use.

The point I am making is that scientists have been their own sternest critics. They have compared different techniques of enquiry, focussed on inconsistencies between the results they provide and gradually found out why these arise. Their method of approach must be contrasted with the blanket statement, still made by some hard-line creationists, that fossils were planted in the rocks when the world was created a few thousand years ago. While it may not be possible to refute this on *logical* grounds, it turns geology into a carefully contrived charade constructed by some super-being for no obvious reason. If one accepts this type of argument then there is no reason to believe anything which one sees around one: it might have been faked by some mischievous spirit just to deceive us. Indeed it is *logically* possible that the world sprang into existence on 1 January 1900 (or any other date). It is easy to make bald statements which predict and explain nothing, but very hard work to build up *detailed* explanations of the natural world. The most severe criticism of extreme religious fundamentalism is not that it is wrong (scientists are also sometimes wrong), but that it discourages people from trying to understand the marvellously complicated world around us.

We describe a beautiful example of the detailed knowledge which proves the extreme antiquity of the Earth. Let us use the word aboriginal to refer to those radioactive isotopes which are not currently being produced by any known process—in other words those which we believe to have been present since the formation of the Earth. If one measures the half-lives of the aboriginal isotopes one finds values which are all greater than 80 million years. These facts, the results of dozens of independent measurements, provide strong evidence that the world has existed long enough, that is well over 80 million years, for all of the many shorter lived isotopes to have decayed to unmeasurably small amounts.

The Mechanisms of Inheritance

We now turn from the proof of the extreme age of the Earth to the question of what controls the forms of individual animals and species. This question is logically independent of any theory of evolution, since there would have to be something which ensures that the offspring of dogs are dogs and of rabbits are rabbits, even if these species had been created in exactly their present forms. Also there must be something which ensures that my children resemble me more closely than they resemble another randomly chosen person.

Our present theory of inheritance is not contained in Darwin's *The Origin of Species*. Indeed his speculations about the physiological mechanisms of inheritance were either vague or wrong. It is often said that Darwin refuted Lamarck's theory that acquired characteristics could be transmitted to descendants, but this is not true. In later editions Darwin increasingly supported Lamarck's theory. Let me quote from the preface to the second edition of his

The Descent of Man, published in 1873, when he had had fourteen years to consider further evidence and correspondence:

> *I may take this opportunity of remarking that my critics frequently assume that I attribute all changes of corporeal structure and mental power exclusively to the natural selection of such variations as are often called spontaneous; whereas, even in the first edition of The Origin of Species, I distinctly stated that great weight must be attributed to the inherited effects of use and disuse, with respect both to the body and mind.*

Indeed in *The Variation of Animals and Plants under Domestication*, published in 1868, he described an ingenious mechanism by which acquired characteristics could be inherited. It involved the migration of myriads of tiny 'gemmules' (one imagines these to be the size of viruses) from the various organs of the body to the germ cells, from which they are transmitted to the next generation. The theory is rarely mentioned today because it turned out to be wholly without factual support. This is not the only case in which people neglect to mention matters which would lessen the reputation of those we revere as geniuses.

Lamarck's theory was first challenged experimentally by Weismann late in the nineteenth century. He systematically cut off the tails of mice over many generations and found that the tails of their descendants were not affected in any way. In fact these experiments were hardly necessary, as a little thought about the foreskins of Jewish men demonstrates. Many later claims to have found evidence for the inheritance of acquired characteristics proved to be wrong, but the abandonment of Lamarck's theory was eventually forced by the rise of genetics.

The key to the mechanism for the inheritance of characteristics was found by Mendel. His experimental research on peas introduced the notion of discrete genes which control individual characteristics and which are transmitted during reproduction. It was not published until 1866, seven years after Darwin's *Origin* and remained obscure until it was revived/rediscovered in 1900. Even then the nature of genes was shrouded in mystery: the detailed molecular structure of DNA and the mechanism by which it encoded genetic information was only discovered in 1953 by Crick and Watson, supported by Franklin and Wilkins, both at King's College, London. The central contributions of Rosalind Franklin to the discovery are only recently being recognized, but her gender was only one of the factors involved in her previous relative obscurity. Another is that she had already died when the other three were awarded their Nobel Prizes in 1962.

Put briefly, a DNA molecule is a very tall stack of almost flat subunits of two types, which may be called AT and CG. AT is actually an adenine-thymine base pair, while CG is a cytosine-guanine base pair, but the details do not concern us. The information in DNA is encoded by the order in which the AT and CG subunits occur in the stack, which is arbitrary. Interestingly the subunits are read three at a time when DNA is used by the cell to synthesize proteins. This involves the production of RNA, followed by a complicated series of further processes which we will not even attempt to describe. DNA has precisely the same status as a floppy disc in a computer. It is completely useless except in the right environment, but can enable the cell to carry out procedures which would be impossible without it.

These flat subunits can be seen in Irving Geis's picture in figure 8.3. They are surrounded by two outer spiral backbones, which act as scaffolding holding the molecule together. The actual molecule is thousands of times longer than the small fragment drawn.

The genotype of an organism is defined as its collection of genes. Most genetic material, that is DNA, resides in the chromosomes within the nucleus of each cell. But there is also extra-nuclear DNA in plasmids and mitochondria, small structures found within cells. Moreover plasmids can transfer DNA between species of bacteria, this being the mechanism for the transfer of drug resistance between bacteria. As the genetic structure of more and more species

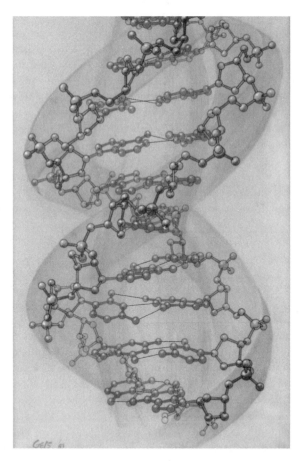

Fig. 8.3 A Small Part of a DNA Molecule
Illustration by Irving Geis. Rights owned by Howard Hughes Medical Institute. Not to be reproduced without permission

have been investigated, it has become clear that the transfer of genes between species has been quite common in evolutionary history. It is commercially important at the present time because of the speed at which genes are being transferred from GM crops to wild hybrids. The notion that different species are rigidly distinct from each other is, quite simply, dead.

The phenotype of an organism is not as simply defined as its genotype, but combines the physical structure of its body and innate behaviour (for example the migration of certain birds and their nest building). The genotype is one of the important factors determining the phenotype, but the environment, including social interactions in some species, is another. It is well known, for example, that a mother's health and diet during pregnancy have a strong effect on the health of her offspring, and therefore of its likelihood of reproducing.

The genotype affects the phenotype in an extremely indirect fashion, since genes actually encode for the production of proteins. At present we understand little about how a variation in the form of certain proteins can lead to a change in bodily form or behaviour. One certainly should not expect to find 'a gene for X': the situation is commonly far more complicated. A physical characteristic may be affected by very many genes, and in the reverse direction a gene may have several different effects. It is well known that the gene which gives some protection against malaria when present singly also leads to sickle cell anaemia for those who possess two copies. While the colour of our eyes is under the control of a very small number of genes, most characteristics, such as our height, are affected by a large number of different genes, as well as by nutrition and disease. Many diseases have a genetic component, but the relevant gene or genes may well have some other effect in people without the disease. It is likely that some diseases are caused by several different genes or combinations of genes. In such cases a drug which cures some people might be completely ineffective against other people suffering from the same disease.

One of the main differences between ourselves and chimpanzees is that only we have an inherited ability to use language. This must be an consequence of the 1.5% difference between our gene pool and that of chimpanzees. A part of the reason is the anatomy of our larynxes, but this is far from being the whole story. How the possession of certain genes, and therefore the ability to synthesize certain proteins, led to our ability to produce grammatical sentences is totally mysterious. Some so far unidentified proteins affect the development of embryos, leading to the appearance of special circuits in the brain, which then enable us to develop language. But no steps in this process are yet worked out, and the enormous progress being made in determining the human gene sequence will not lead to a quick solution to this problem.

The precise definition of phenotype involves further complexities once one starts looking at individual species, and Richard Dawkins has argued that the notion needs to be extended well beyond its traditional scope. If one accepts that the shell of a snail is a part of its phenotype then one should admit the web of a spider in the same way, or at least the genetically programmed tendency of spiders to spin webs of particular designs. The webs and the shells are both subject to selection forces in exactly the same way as the bodies of these animals. Dawkins argues that the same idea should be applied to the dams of beavers and the mounds of termites. In fact Darwin made a related point when referring to cooperative behaviour in human tribes.

More controversially, Dawkins has argued that one should think of selection as acting on gene pools rather than on individual organisms. It is not even clear what constitutes an individual for species such as slime molds, whose life cycle includes a stage similar to a single-celled amoeba and a later one in which groups of cells come together to produce a larger organism. A similar difficulty applies to plants such as bamboos, which reproduce mainly by sending out horizontal underground stems called rhizomes from which new shoots develop. According to one definition single bamboo individuals may cover several hectares! Extreme examples should not, however, be taken as proofs that the study of the survival and propagation of individuals should be consigned to the dustbin. The genotype and phenotype are so deeply intertwined that neither can be regarded as a secondary issue if one wishes to understand evolution fully.

Theories of Evolution

A theory of evolution differs from what we discussed in the last section by asking how species came to exist in their present form. This is a historical question about events which happened long ago, and for which most of the evidence has disappeared. In addition the variety of forms of life is so great that there is little hope of providing a single account of how evolution occurred. The relationships between species include:

- predator–prey, e.g. tigers eating ruminants.
- host–parasite, e.g. cuckoos, tapeworms, leeches, and viruses parasitizing their hosts.
- slavery, e.g. of one species of ant by another.
- symbiosis—the intimate physical association of two species to their mutual benefit—possibly including mitochondria and chloroplasts, now essential parts of the cells of animals and plants.
- interspecies cooperation, e.g. when flowering plants are dependent on a particular species of insect for pollination.
- mimicry, e.g. when a harmless insect mimics a poisonous one in order to reduce its own chances of being eaten.

The more one learns about this subject, the more one realizes that Nature has tried a vast range of different ideas. The subject is messy in a way that physical scientists and mathematicians can hardly imagine. This does not mean that it is unscientific, but rather that one should not expect that the final conclusion will be a set of laws which can be written down on a few sheets of paper. Instead one is aiming towards a vast collection of case studies involving transferable ideas and techniques. In contrast to the matters discussed so far, we will see that there is still plenty of controversy about the theory of evolution. This is not surprising, since the evidence is still coming in.

Darwin conceived of his theory of evolution in 1838 shortly after returning from his famous voyage in the Beagle to South America, and in particular to the Galápagos Islands, between 1831–6. Plagued by ill-health, he soon retreated to Down House in Kent, where he devoted the next twenty years to writing on a variety of topics in biology, and to preparing a massive tome on his theory of evolution. He was in no hurry to complete this work, and might never have done so if he had not received a crucial letter in 1858. This came from Alfred Russel Wallace, and presented very similar (but less well worked out) ideas, asking Darwin to assist in getting it published! In the end their theories were published jointly in the same Proceedings of the Linnean Society in 1858. Rather surprisingly, in view of what happened a year later, there was almost no public reaction. Following this shock, Darwin plunged into a frenzy of activity to get a reduced version of his planned book published as quickly as possible. It appeared in 1859 as *The Origin of Species*, and was an instant best-seller. Over the next few years there was intense public debate about the scientific merits of the theory, and about its religious implications.

In *The Origin* Darwin discussed in detail evidence for the evolution of a variety of wild and domestic species. His originality lay in proposing a particular mechanism for evolution and in describing a very wide range of evidence relating to his ideas. The key idea in his theory was that individuals vary slightly from each other and are engaged in a constant struggle to survive with predators, parasites, and each other in a constantly changing environment. The intensity of this struggle is something we humans are hardly aware of; only a small percentage of most organisms live long enough to reproduce and for many species the proportion is *far* smaller. In these circumstances those individuals which survive to reproduce pass on their characteristics to their descendents. Darwin argued that this leads to slow but progressive changes in species which, given enough time, can totally transform them. His central idea is summed up in the phrase 'evolution by natural selection'.

One of the key issues for Darwin was to emphasize the lack of clear boundaries between species. For those who took the Bible literally, or their modern-day equivalents, species were created individually by God in the form he wished them to have, and therefore would not vary over time. Evolution, however, supposes that forms of life change gradually over time, with the inevitable consequence that if one could take some creature and a sufficiently distant descendant, everyone would agree that they were different species, while at some intermediate time people would not be able to agree whether they should be regarded as different species or not.

Darwin's theory proved very controversial, because it was completely amoral and contrary to the teachings of the Bible. The phrase 'the survival of the fittest', has often been ridiculed as a tautology, and was first used by Spencer in 1852 in an essay which came close to formulating the general principle of natural selection. It was only adopted by Darwin in 1866 at the urging of Wallace. Dawkins devoted a whole chapter of *The Extended Phenotype* to the discussion of this unfortunate aphorism, quoting the letter from Wallace to Darwin and explaining how to define fitness in a non-tautologous manner.

A serious weakness of the concept of fitness is that it depends heavily upon the climate, which may vary irregularly from year to year. What aids survival one year may not do so the next. Recent evidence that evolution follows changes in the climate extremely rapidly has been provided by Peter and Rosemary Grant's studies of 'Darwin's finches' on Daphne Major in the Galápagos islands.[5] They have followed the lives of every single finch on this tiny isolated island for about twenty years, recording the beak sizes and other details of each bird, as well as its breeding success. The large variations between the beaks of the thirteen different species are crucial in determining which of the different types of seed and other food each species is best adapted to eating: even tiny differences can affect the survival of an individual greatly. The study provides convincing evidence that measurable inherited changes in the populations can occur even in a single generation, when failure of the annual rainfall causes high and *selective* mortality. Daphne Major is ideal for this study because the climate is subject to extreme droughts and floods, depending on El Niño.

On a much longer time scale the development and retention of major organs depends upon the environment. Thus the evolution of wings might have been an advantage at one time to enable some animals to escape from predators. At a later time the degeneration of the wings of emus was also an advantage, because of the lack of predators in their habitat in Australia and the high energy expenditure involved in flying. In Mauritius the dodo's loss of wings was first an evolutionary advantage, and then much later a fatal disadvantage. Their inability to fly led to their extinction because of the appearance of new predators—humans and species introduced by them.

The three great mass extinctions of the last billion years also demonstrate the limited usefulness of the concept of fitness. Sixty five million years ago dinosaurs had been the dominant land animal for over a hundred million years. Mammals had existed for a similar period, but were a class of marginal significance. Within a few million years of the mass extinction which happened at that time, dinosaurs had become extinct, while mammals had diversified and increased in size dramatically. Mammals were obviously better able to survive whatever eliminated the dinosaurs. Equally obviously their 'fitness' in this respect was not a result of their having evolved to cope with a type of event which only occurs every hundred million years or so.

Darwin's theory of natural selection was substantially influenced by Malthus, who described the consequences of the geometric increase of a reproducing population if unconstrained by any external factors. The force of this argument is easy to demonstrate with some simple arithmetic. Consider a population of a million short-lived insects, each of which produces a large number of offspring every year out of which on average only one survives. The population is then stable. Now suppose 1.01 survive (on average of course!); then after 1500 years the population will have grown to about 3 trillion insects. Of course this is not possible, and in reality some factor limiting the increase will come into effect. On the other hand if only 0.99 survive on average then the population will disappear entirely after 1500 years. So a tiny change in survival rates has a catastrophic effect on the population within a period of time which is invisibly short by comparison with the total age of the Earth. When one contemplates these figures the immediate question is how populations remain stable at all. There are many possible answers. One is that as the number of insects rises the number of birds also rises because the birds have more food, with the result that the population is kept in check. Another possibility is that the number of insects does in fact explode but it eventually runs out of food and collapses again. The sporadic plagues of locusts in Africa and elsewhere show that such events can have lifetimes measured in years rather than thousands of years.

It is sometimes said that almost all mutations are deleterious, so evolution cannot lead to positive effects on the species, however long one waits. This argument is simply wrong. A mutation which reduces the chances that an individual will reproduce is rapidly eliminated from the gene pool. Its only influence is to reduce the population slightly, but even this effect is negligible if the food supply is the main factor holding the population in check. Even if such mutations occur frequently they have no cumulative effect. On the other hand, those rare mutations which improve the survival rate of the offspring possessing them will become more common, unless chance factors eliminate them at an early stage after their appearance. Even if they confer only a small advantage they may easily become dominant within a thousand years in a small population.

Over a period of ten thousand years ten such changes could have accumulated, even if only one feature was selected at a time. In reality one might expect many times that number of variations to have disappeared or become dominant. The extent to which a species changes as a whole depends upon whether the variations are mostly in the same direction. This in turn depends upon whether the climate steadily changes with time and whether competitors move into the area or become extinct over this period. Certainly change is not inevitable: cockroaches have survived as an order of insects almost unchanged for over 300 million years. While they might therefore be described as primitive, there is no law that primitive organisms must change if they are so well adapted to their habitat that change is not necessary. Other examples include ferns, already well established by 250 million years ago and not eliminated by the 'more advanced' flowering plants which appeared a hundred million years later.

There is recent evidence that under favourable circumstances evolution can indeed be very rapid. Lake Victoria in Africa possesses more than two hundred species of cichlid fishes, much more closely related to each other than to the cichlid fishes of Lake Malawi, for example. It has been know for some time that these must have evolved over a period of less than a million years, but everybody was astonished by the results of a survey of the lake published in 1996. Cores taken from the deepest point of the lake showed that 12,400 years ago it was completely dry: remains of grasses exist and can be dated reliably. The conclusion is inevitable. All of these species evolved from a few ancestors since that date![6]

One and a half centuries after Darwin the evolutionary record is still too incomplete to trace the evolution of many species, but much progress has been made. One of the well known pieces of evidence for actual evolution is that of the horse. The early equid Hyracotherium, which appeared about 50 million years ago, was so different from the present form that its fossils were not immediately recognized as being related to modern horses. It was about 50 centimetres in height, had three functional hooves on each foot, lacked the muzzle of our horse and had substantially different teeth, not well suited to grazing on grass. The fossil record does not show a directed development from the 'primitive' Hyracotherium to the modern horse. Many changes of direction and different lines developed and survived for millions of years. Some changes, such as that to a grazing type of dentition, happened rather suddenly (in evolutionary terms), probably in response to a change in climate and the wider distribution of grasses. Nevertheless the fossil record is sufficiently complete that one can be confident that the modern horse is indeed a descendent of Hyracotherium.

Among the most convincing chapters in *The Origin of Species* are the two on the geographical distribution of species. Darwin explained with great clarity why one might expect land or sea barriers to lead to separate evolution of species, and provided a wealth of detailed observations to support this. For example:

> *Turning to the sea, we find the same law. No two marine faunas are more distinct, with hardly a fish, shell, or crab in common, than those of the eastern and western shores of South and Central America; yet these great faunas are separated only by the narrow, but impassable, isthmus of Panama. . . . [On the other hand] many shells are common to the eastern islands of the Pacific and the eastern shores of Africa, on almost exactly opposite meridians of longitude.*

I cannot reproduce these chapters *in toto*, but strongly urge readers who have not yet done so to read this wonderful book for themselves.

We have seen that the geological record establishes the existence of evolution in the sense that species appear, change in form, and become extinct over long enough periods of time. The remaining issue is whether Darwin's mechanism for evolution is the dominant one. One of the earliest serious criticisms of Darwin's theory was made by Jenkin in 1867. He wrote that Darwin's theory depended on the *assumption* that there is no typical or average animal, no sphere of variation, with centre and limits; the theory could not, therefore, also be used to prove that assumption. His opposing view was that of a race maintained by a continual force in an abnormal condition, and returning to that condition as soon as the force is removed. Darwin was quite troubled by Jenkin's comments, since his own book recognized the phenomenon of reversion, by which occasional individuals within a domesticated variety changed back towards the ancestral form after having been bred so as to have quite different characteristics. The trite answer to this objection is that effects observed when breeding domestic species over a few hundred years may look very different from a perspective of several million years. The difficulty is finding evidence one way or the other.

The effect of a change of perspective can be dramatic. The area of Dulwich in inner London was almost the same twenty years ago as it is today. The most obvious changes have been the disappearance of sparrows, the appearance of a few new primary schools, and the proliferation of loft extensions. If one travels back in time two hundred years, almost all of the buildings have disappeared and Dulwich was a tiny village well outside London, notable only for its association with the Elizabethan actor Edward Alleyn, who founded Dulwich College there in 1619. Turn the clock back by two thousand years, just before the Romans arrived, and hardly anything remains of London itself. But it is clear that every change in the environment was a small one, and every building took months to erect. If this can happen in such a period, how much more might change in a million years?

Jenkin also made another objection to Darwin's mechanism for evolution, which is rather more technical. If a member of an abundant species has a modification which renders it only slightly more fit, then the variation will disappear by blending with the rest of the population before it has become widespread. This dilution argument made evolution difficult to understand. Darwin was unable to answer this criticism since he was not aware that evolution might be controlled by discrete genes which do not disappear by blending, but only increase in frequency or disappear.

Major efforts to find evidence for evolution by the statistical analysis of populations were made by Weldon, Pearson, Bumpus, and others around the turn of the twentieth century. Eventually this enterprise receded into the background, because the discovery of genes provided the mechanism for inheritance and for the progressive modification of species. This, more than anything, settled the issue for many scientists. Nevertheless controversies have continued within the field. Some evolutionary biologists, such as Dawkins, maintain that organisms should be understood primarily as machines for the transmission of genes, which are the key objects worthy of biological study. In his book *Lifelines*, Steven Rose criticizes this kind of ultra-Darwinism, pointing out that Mendel's observation of characteristics of peas determined by single genes is far from typical of their mode of action. Different cells in the body of an animal or plant have a wide variety of forms because different genes are expressed (switched on or off) in them. Control over gene expression is not only a method by which the parts of a growing organism differentiate. It is becoming increasingly obvious that cells may switch particular genes on or off for a variety of reasons—for example as a method of defending themselves against viral invasions. The expression of genes therefore depends on the extracellular environment and even the environment outside the organism. The picture which Rose paints involves such strong interactions between genes, organisms and their environments that nothing worthwhile can be said without taking all three into account.

There are other sources of disagreement in the scientific community. Darwin claimed that evolution occurs by the accumulation of very small changes over many millennia. Thomas Huxley, one of Darwin's strongest supporters, wrote to him in 1859 to say that he had loaded himself with an unnecessary difficulty in adopting 'Natura non facit saltum'[7] so unreservedly. More recently Gould, Eldridge, and other punctuationists favour occasional major changes which are very rapid *if one measures on the scale of millions of years*. The fossil record provides some support for punctuationism but is so incomplete that it is impossible to come to a final decision about it at the present time. Indeed it is not clear what counts as a small change, since it is now known that the alteration of a single base pair in DNA can have visible and important effects on the whole organism. Nor is it clear that the thesis of the punctuationists is as radical as they have claimed. Darwin himself accepted the possibility of long periods of stasis interspersed with shorter periods of more rapid change.

One of the preferred mechanisms for rapid change goes by the unattractive name of allopatric speciation. The idea is that a small group in some species is physically isolated in a region to which they are not well adapted, possibly because of a sudden change of climate. Over a period of perhaps tens of thousands of years they evolve rapidly before rejoining the main population, which they may then drive into extinction. If this happens, the fossil record would show the instantaneous appearance of a new species and the disappearance of the old. Another possibility is that the two species occupy sufficiently different ecological niches that they both continue in existence. In this case the new species might seem to have appeared out of nothing. The first stage of this process is exactly what has been observed with the cichlid fishes of Lake Victoria. It is not surprising that there is little evidence of similar examples in the past. The transitional forms are supposed to be small populations because this makes the inertia preventing rapid evolution much weaker, but this very fact implies it is unlikely that any transitional fossils would ever be found.

There are several forms of contemporary evidence for the importance of natural selection in micro-evolution (the development of relatively minor changes). The most obvious is the development of drug resistance by a variety of different pathogenic micro-organisms, such as the tuberculosis bacillus, which now once again poses a real threat to world health. Since the introduction of the first antibiotics during the Second World War, they have been used steadily more widely as if they were magic bullets which would solve all problems. Within the last twenty years more and more microbes have started developing serious levels of drug resistance, a fact which is explained in terms of their evolving in the face of what appears to them to be a more hostile environment. Another example of such biochemical evolution is the development of resistance to warfarin by rats. The use of warfarin started in the UK in 1953, and by 1958 the first populations showing resistance to its effects had already appeared. The fact that they spread from a few small isolated locations suggests that the genetic changes happened in single individuals which then passed on the resistance to their descendants.

Macro-evolution (for example the appearance of entirely new organs) is harder to document, but not impossible. Over the last fifteen years an enormous amount has been discovered about the evolution of whales from their land-living ancestors fifty million years ago. The same can be said about pterosaurs, flying reptiles which lived mainly during the Jurassic period.[8] I will not repeat what has already been written about many times.

Science has at last got to the point at which it might be possible to understand the genetic basis for the development of the eyes of flat fish. These bottom-living fish have such distorted heads that it is difficult to imagine how any being with a sense of beauty or design could have created them. Can science do any better? There are about three hundred species of flounder, some of which have both of their eyes on the left and some on the right of their bodies. From our point of view their eyes are on the tops of their heads, because they habitually swim on their sides on the bottom of the sea. The young are born symmetrical, but after a short period undergo a metamorphosis in which one of the eyes starts to migrate rapidly to the other side of the head. During the metamorphosis there are major changes in the bones, nerves, and muscles enabling the eye to move and then function in its new position.[9] It would be extremely interesting to determine the precise causes of this major change of form. It is known that there is a sharp increase in the production of thyroid hormone at the time of metamorphosis, and this probably plays an important role in the development of the asymmetry. Once it had appeared it is easy to see how selective pressures would have preserved it as an adaptation to life on the sea bottom. Unfortunately determining the genetic mechanisms involved has had low priority because of its lack of glamour or obvious commercial relevance. From a Darwinian point of view the interesting question is whether the movement of one eye to the 'wrong' side of the head was the result of the accumulation of a large number of small genetic changes, or whether it depended upon a single gene.

Although the last question remains unanswered, there has recently been a breakthrough in the understanding of the formation of limbs in arthropods (animals with segmented bodies and external skeletons), and particularly insects.[10] Evidence that arthropods evolved from simpler ancestors something

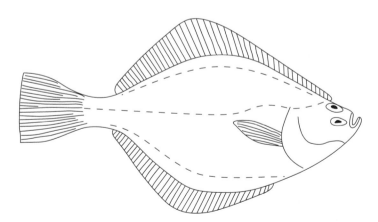

Fig. 8.4 An Adult Flounder

like today's annelid worms comes from a vast fossil record, going back over 500 million years.[11] The details of how this happened was, until recently, a mystery. By comparing the precise structures of certain genes and proteins in very different species, scientists have now found the genetic mechanism which suppresses the formation of limbs on most segments of the bodies of insects such as Drosophila. Unsurprisingly the mechanism is very complicated, and may well need revision in detail, but it is clear that such questions can now be answered with sufficient effort. We are close to the point at which we can identify the precise features of their DNA which cause insects to have six legs and spiders eight!

Some Common Objections

The claim that evolutionary theory cannot be correct is often repeated, mostly by people who have a religious agenda. Their objections should nevertheless be addressed. Several arguments have been put forward. One is that the evolution of an organ such as the eye is inconceivable, since each of its parts depends so intimately upon the others for its correct functioning. It is claimed that the eye is as obviously designed as is a camera, and, wherever there is design, there must be a designer.

Unfortunately (for its proponents!) this argument is misconceived. It would only carry any weight if any intermediate but less developed organ would be useless, since evolution is required to proceed by the accumulation of small changes, each of which has direct evolutionary advantages in itself. Darwin himself discussed this issue and provided examples to prove that several intermediate stages in the development of the eye do in fact exist in various animals. Computer models have also been constructed which show how the eye could have evolved in accordance with Darwin's theory. This does not of course prove that it did evolve this way, but the argument that it could not have done so is false. The suggestion that intermediate stages in the evolution of the eye would be useless is also obviously false. Many people have various degrees of short and long sight, but it is quite clear that even extremely poor sight is vastly better than no sight at all. Moreover, the better the performance of one's eyes, the better one is able to find food and to avoid threats such as poisonous snakes. An eye without a fovea (the central area of the retina which provides particularly sharp vision) would be entirely functional but any concentration of rods and cones there would have clear advantages. There are therefore large benefits obtained from extremely poor sight as well as great evolutionary pressures for poorly functioning eyes to improve.

A second argument against Darwin is that his theory was conceived in an extreme capitalist society. It was accepted, either consciously or unconsciously, because it provided a justification for the exploitation of weaker members of society. Darwin himself was largely apolitical. He abandoned traditional Christianity fairly early and became an agnostic, but took some trouble not to press his views in public beyond what was needed to support his scientific theories. On the other hand Malthus's doctrine was indeed frequently used to justify the brutal suppression of the lower classes and various minorities. One cannot dispute these facts, but the reasons why we accept or reject Darwin's theory now need not be exactly the same as those which may have made it attractive in the middle of the nineteenth century. Each generation re-examines a theory for weaknesses and strengths, and over a long enough period one finds out which parts of it retain their validity. The 'capitalist objection' was responsible for the rejection of Darwin's theory in communist Russia in favour of a Lamarckian evolutionary theory. This had the political merits of permitting the skills acquired by an individual's efforts to be passed on to his/her descendants. Unfortunately science does not bend itself to political wishes, and the outlawing of standard genetics in the Soviet Union under Stalin by his head of agriculture, Lysenko, proved disastrous for Soviet agriculture.

The idea that evolutionary theory justifies the exploitation of the weak depends on the underlying assumption that the way things have happened is by definition 'good' and that opposition to it is misguided. Taking this view is, however, a matter of choice and not of fact. One can as easily argue that our moral task is to provide help for those whose lives are less than full for genetic, environmental, or even accidental reasons. A similar mistaken argument is still being actively used to justify discrimination against homosexuals and women: namely that what the discriminator considers to be 'natural' is also by that fact morally right. This ignores the obvious fact that the human species has only become what it is by trying to improve on what it was given by nature. If we stuck to what was natural civilization would never have arisen. A defence of some moral principle by appealing to evolutionary theory is what philosophers call a category mistake: facts cannot dictate ethics. Arguing this way is almost always an excuse for promoting the interests of the social or ethnic group the person concerned happens to belong to, conveniently identified as the superior race (or gender). In nineteenth century Europe the 'naturally superior' race were, unsurprisingly, white European males.

The above ideas are not taken from a recent revisionist interpretation of Darwinism. In fact Darwin's greatest supporter, Thomas Huxley, expressed similar ideas with great eloquence in his 1893 Romanes Lecture *Evolution and Ethics*:

> *There is another fallacy which appears to me to pervade the so-called 'ethics of evolution'. It is the notion that because, on the whole, animals and plants have advanced in perfection of organization by means of the struggle for existence and the consequent 'survival of the fittest'; therefore men in society, men as ethical beings, must look to the same process to help them towards perfection. I suspect that this fallacy has arisen out of the unfortunate ambiguity of the phrase 'survival of the fittest'. 'Fittest' has a connotation of 'best'; and about 'best' there hangs a moral flavour. In cosmic nature, however, what is 'fittest' depends upon the conditions . . . As I have already urged, the practice of that*

*which is ethically best—what we call goodness or virtue—involves a course
of conduct which, in all respects, is opposed to that which leads to success in
the cosmic struggle for existence . . . Let us understand, once for all, that the
ethical progress of society depends, not upon imitating the cosmic process,
still less running away from it, but in combating it.*[12]

There is a group of people, already mentioned in the Gallup poll, who
accept evolution over a very long time scale, but believe that it took place under
God's guidance, with the intention of leading eventually to the appearance of
human beings. This is not what Darwin proposed in his book, but it is a long
established idea, and should not be condemned on those grounds alone. It is not
tenable if one believes that natural laws completely determine everything about
the world, as many scientists do. However, the position of this book is that the
natural laws which we know represent our best attempt to understand the world,
and may only succeed to a limited extent. Miller has given a lengthy defence
of the idea that the world can be wholly governed by natural laws while being
simultaneously being subject to God's continuous guidance and direction.[13]

Let us examine how this question might in principle be resolved. Suppose
that substantial numbers of well preserved fossil remains of precursors of Homo
sapiens were eventually discovered, providing an essentially continuous record
of our evolution over the last five million years. Suppose also that a detailed
examination of the DNA of ourselves and the other great apes allowed us to
reconstruct the entire genome of our common ancestor five million years ago
and the probable sequence of stages by which our DNA changed over that
period to its present form. The question would still remain: why did this particular sequence of DNA changes occur out of a wide range of other unknown
possibilities which might have led to 'just another great ape'? An answer to this
question would have to depend on the detailed historical and environmental
origins of our own species, information which seems unlikely ever to be available in the required detail. It is undeniable that the other great apes have not
followed our route to language and sophisticated tool-making, and only in the
last half million years does it begin to seem inevitable that we would develop
the way we have.

The above programme could run into difficulties. Perhaps nobody will be
able to find a plausible sequence of small changes of our DNA over the period
of five million years. Every such sequence might contain a form which would
be less 'fit' in the evolutionary sense than the previous one in any imaginable
environment. This would be a truly serious development, which would inevitably be taken by the advocates of guidance as a proof of divine involvement in
our appearance. Evolutionists would no doubt regard it as just another challenge
to their ingenuity. On the other hand if the programme were successful, advocates of guidance could still say that our appearance is such a singular event that
without a proof of its *inevitability* they still felt fully justified in regarding the
evolution of our species as directed.

The fact that this position is logically unassailable does not prove its correctness: the existence of poltergeists or leprechauns is not logically impossible, just *very* unlikely. Guided evolution is more plausible than either of these, but there is no compelling *evidence* for it which would be recognized by a person who does not already have the relevant religious convictions. Science has progressed as far as it has by looking for natural explanations of the world, and appeals to divine guidance inevitably discourage people from continuing the search. The attempt to explain our appearance along naturalistic lines might eventually fail, but much more has been learned by trying than ever could have been by accepting passively a 'solution' based on no evidence.

There are other issues which the advocates of guidance need to address. One is that the most important evolutionary developments took place long before humans appeared on the scene. There are many fossils of vertebrates dating back to 400 million years ago, and the first very primitive mammals had already appeared by 200 million years ago (during the Triassic period). So eyes, skeletons, digestive systems, blood circulation, spinal cords, and brains all existed at that time, and human beings are merely one of a very large number of variations on a well tried body scheme. If external guidance was indeed needed it was mainly during this distant era, not in the few million years when humans evolved. There are those who consider that the rapid development of complex organisms in the late Precambrian is incapable of explanation and is therefore evidence of divine intervention. The evidence this far back is unfortunately so fragmentary that no conclusions can be reached either way.

A problem for those who believe that guidance was involved in the evolution of human beings is the failure of the guiding spirit to create as good a product as it could have. Those who feel convinced by arguments from design need to consider the following facts.

- Humans often develop impacted wisdom teeth because our jaws are not big enough to accommodate the number of teeth we have.
- Our appendixes have no positive function, but occasionally cause acute illness, which results in death if rapid medical intervention does not occur.
- We are peculiarly liable to choking on food, and indeed occasionally dying, because of the structure of our throats, which are designed differently from those of other mammals.
- Our pelvic anatomy makes childbirth painful for many women, and resulted in many deaths until very recently. Once again this is not a common problem for other mammals, and is connected with our bipedal locomotion and large heads.
- We have a blind spot in our visual field because of the peculiar manner in which the neural pathways pass in front of the light-sensitive receptors on their way to the brain.
- Humans suffer from the disease scurvy when our diet is deficient in vitamin C. Many other vertebrates and plants can synthesize this vitamin themselves.

It is very easy to understand how such things might have arisen if the process of evolution had no overall direction, but just consisted of responses to selective pressures.

We have concentrated above on humans, but objections to guided evolution can also be found in the fossil record. If a creator has been directing evolution towards the currently existing forms of horse, why would he/she produce a large variety of other lines of evolution from Hyracotherium, almost all of which subsequently became extinct? Why produce the huge variety of dinosaurs, all of which became extinct by around 65 million years ago, except for the small group which may have developed into modern birds? The case of trilobites, a group of arthropods, is even more extreme in that, after appearing about 540 million years ago, they became one of the dominant marine animals, only to become entirely extinct by 250 million years ago. All these facts (any many more like them) make the psychology of a guiding creator difficult to comprehend.

A typical answer to such questions is that we cannot understand God's purposes because of our limited perspective and finite intellect. Nobody would deny that our intellect is finite, but our self-imposed task is to try to find explanations for the way the world is, and not to accept mysticism as an 'explanation', when in fact it explains nothing. Another response is to refer to original sin as the cause of our present suffering. I have never understood this doctrine, and feel grateful not to have been brought up in an atmosphere in which people are supposed to acquire guilt/sin by association with the deeds of distant ancestors. There is indeed plenty of human evil in the present world, but this does not explain a substantial proportion of the causes of human suffering.

I should not end this section without acknowledging the existence of people such as the American philosopher Michael Ruse, who has written many books exploring whether the gulf between Darwinism and Christianity is really as sharp as is often considered.[14] Starting from a Quaker background, he has argued that most of the difficulties of Christianity pre-existed Darwinism. Darwinism may explain why infectious diseases exist, but the question of God's ultimate responsibility for such things already exercised people's minds long before the nineteenth century. God's best defence is that even he is constrained by the requirement of logical consistency. Our freedom to choose between better and worse (good and evil) implies that there must exist things which are worse, and in a sufficiently varied world there must be many things which are worse and which are not easily evaded. For most people a world without infectious diseases would be a better one, but it would also be a world in which mass starvation was much more common. As long as people breed faster than the food supply can support, there must be a mechanism for removing the 'excess' members of the population. There is no possibility that this will appear morally good to its victims. On the other hand a world in which the birth rate was exactly adapted to the food supply would be a world in which the freedom to decide how many children one had did not exist. Such agonizing dilemmas confront people responsible for guiding social policy, as well as those who seek to understand the mentality of their God.

Discussion

In earlier chapters I have argued that much of our knowledge of physics is provisional, and that seemingly impregnable theories have subsequently turned out not to be correct in any absolute sense. It therefore seems inconsistent to suggest that in biology and geology it is possible to acquire knowledge which will never be proved mistaken. Nevertheless, it is certain that the Earth is a few billion years old, and that continents have moved over the surface of the Earth during the last several hundred million years by a mechanism called plate tectonics. It is also established that information about the form of living creatures is carried by their genes in the structure of their DNA, that animals and plants evolve over long periods of time, and that most species which have ever existed are now extinct. The reason for being confident that the above statements (and many more which I have no space to list) are *certain* is that the evidence for them comes from so many independent sources, which all corroborate each other. Moreover the statements do not depend upon the abstractions of mathematics in order to understand them. Mathematics is a wonderful tool, but history has shown that theories which can only be expressed in mathematical form are liable to radical change with the passage of time. Of course nothing can be certain in an ultimate sense, but my confidence that the above statements will still be believed in a thousand years time is much greater than my belief that quantum mechanics or general relativity will be remembered by most scientists at that time. I would be happy to be proved wrong, because I find the changes in our views about the nature of the world fascinating, but I do not expect this to happen with respect to the above facts.

Darwin's proposed mechanism for evolution needs to be separated into two parts. One can hardly dispute his claim that those organisms which have more offspring will also pass on their peculiar characteristics to a greater extent than those which have fewer offspring or even die before reproduction. On the other hand his view that evolution only proceeds by the accumulation of individually small changes is more controversial. There are many serious scientists who do not believe this, and the evolutionary record remains too fragmentary for a final view to be possible.

The study of evolution is quite different from physics. Biologists cannot hope for a single coherent explanation of all phenomena in their subject because Nature is far too varied. Excluding such generalities as the fact that living organisms all contain carbon, oxygen and hydrogen, every biological law has exceptions. The following list of items are a selection from those which interest evolutionary scientists, and several might be regarded as statements of fact. They are *not* articles of faith, but ideas which have so far been steadily more strongly confirmed as knowledge accumulates. I have omitted some further items deliberately because they are still controversial. Several of the principles listed below are present either explicitly or implicitly in *The Origin of Species*. Others are completely absent.

- Species are not immutable, but have evolved over long periods of time, and most species have eventually become extinct.
- There is not a sharp boundary between species. Nor is there a sharp division between the notions of species and varieties.
- Organisms have inherited characteristics, and those which survive to reproduce pass on their characteristics to their offspring. Those which don't, don't.
- The evolution of new species or organs cannot involve the existence of intermediate stages which are less well adapted than what they evolved from.
- The origins of the human species go back several million years and are closely tied to those of chimpanzees, gorillas, and orang-utangs.
- Information about the form of living organisms is carried in their genotypes, ultimately encoded in their DNA (or, rarely, their RNA).
- The fossil record and DNA analysis support the idea that all species arose from a single-celled ancestor several billion years ago.
- The relationship between the genotypes and phenotypes of organisms is often extremely complicated, and must be investigated case by case.
- It is essential to distinguish between plausible stories about how evolution might have occurred and testable hypotheses.

I might also have included the controversial

- One should seek explanations for the evolution of living organisms which are not dependent upon design or external guidance.

This should be regarded as a methodological principle, i.e. it describes the method by which science studies all phenomena, not just those relating to evolution. But many scientists believe it to be evident that such explanations exist, and that the only task is to find them. This belief cannot be proved beyond any possibility of challenges, but the way of proving it wrong is by finding a case in which it definitely fails. This has not yet happened.

It is rather implausible to describe such a list of principles as a theory. It is really no more than a summary of a programme which is so vast that it could not be committed to paper: it would consist of all of the knowledge which has accumulated for each of millions of species since the subject began. The idea that there should exist a theory with concisely formulated and verifiable hypothesis is clearly not appropriate in the biological sciences for two reasons. The first is that the variety of life is so vast and unlike the subject matter of the physical sciences that attempts to cast evolution in a rigidly deductive form are not appropriate. The second is that evolution has occurred in response to erratic and occasionally cataclysmic variations of the climate. The real issue is whether it is possible to apply the above methodology to individual species and come to interesting and detailed conclusions about them.

It has been said that evolution is unscientific because no experiment could refute it. It is just a series of tales, which can be elaborated without end whatever new facts are discovered. This criticism is often intended to suggest that if there is not a watertight proof of any of the possible *theories* of evolution, the question of whether *evolution itself* occurred must remain open. This argument is flawed. There are many discoveries which would force a serious re-evaluation of the whole subject, were they to be made. I mention the discovery of stone axes embedded within rocks which can be reliably dated as over 100 million years old; the discovery of a fossil human skeleton within the stomach of a Tyrannosaurus; the discovery of an animal or plant whose DNA was wholly unrelated to that of any other species; the discovery of an animal species less than ten million years old which has two backbones instead of one; the existence of a mammal whose brain was located in its abdomen. *Every* evolutionary theory predicts that no such discovery will be made. The evolution of an entirely new type of animal in a few generations would certainly destroy Darwin's theory, but it would not be a fatal blow to evolution itself.

There is an extraordinary amount of evidence which only makes sense if one accepts that evolution has in fact occurred. Our understanding of it may not be perfect and may need substantial revisions in certain directions, but there is no scientific basis for doubting its essential correctness. Darwin's theory of how evolution occurred provided a spur for the search for a mechanism of inheritance, and after great effort this was found in the existence of genes and the structure of DNA. The idea that evolution is an arbitrary collection of tales only makes sense from the false perspective that finding a consistent *and detailed* explanation for a huge collection of facts is an easy task. In fact it is enormously difficult. Scientists often struggle for decades trying to find an explanation of their experimental or observational data, and experience enormous pleasure from the eventual discovery of a coherent theory. Claiming that evolution is unscientific is very easy, but finding an alternative explanation for the detailed facts which it explains would be much more impressive. Simply saying that everything was created by God as he willed it discourages us from seeking an explanation of a multitude of different facts. In the end one has to decide whether one wants to understand the enormous variety of Nature. If one does then genetics, evolution, and natural selection must be an important part of any detailed explanation. They may form the entire explanation, but this would be difficult to prove. Our scientific understanding of evolution could be wrong if some super-being has deliberately created vast quantities of false evidence for some unknowable reason—but this is true of all human knowledge.

Notes and References

[1] I refer to the period between about 260 and 290 million years ago.

[2] Stomata are pores which allow the movement of gases into and out of leaves. In particular they allow carbon dioxide into the leaves for photosynthesis. See McElwain et al. 1999 for details of this research.

[3] The shaded regions are included to make the boundaries of the present continents clearer; they do not represent ancient seas.

[4] Bradley 1999

[5] Weiner 1994

[6] Johnson et al. 1996, Barlow 2000

[7] Nature does not make jumps.

[8] Miller 1999, p. 264

[9] Okada et al. 2001

[10] See Levine 2002 for references to the original papers in the same issue of *Nature*.

[11] Edgecombe 1998

[12] Huxley 1894, p. 80–83

[13] Miller 1999

[14] Ruse 2001

9
Against Reductionism

Introduction

The extraordinary development of physical science over the last two centuries has led to claims that it is capable of explaining all aspects of reality. Some scientists, mainly those in subjects close to chemistry and physics, have proposed a programme, called (scientific) reductionism, describing how this is to be achieved. While some scientists equate a disagreement with the programme to a rejection of the scientific method, others consider that its programme is absurdly narrow-minded.

Reductionists start by ordering scientific fields according to how fundamental they are. There are several slightly different ways of doing this, but the following will suffice here.

Each of the subjects in the box below is supposed to be fully explicable in terms of the one directly below it. The 'explanation' of any phenomenon in terms of something at a higher level is ruled out as being methodologically unsound or even meaningless, on the grounds that an event cannot have two different and unrelated causes. Thus every type of brain activity is claimed to have a complete explanation in terms of its constituent neurons and the chemicals which affect their behaviour. Consciousness must be explained in terms of brain physiology if it exists at all. Many reductionists reject the existence of souls in the strongest possible terms. They are confident that their thesis will ultimately be accepted,

social structures
consciousness
brain physiology
the biology of cells
molecular biology
chemistry
physics
theory of everything

and believe that the role of the scientist is to fill in the details of how the reduction is effected.

One may contrast reductionists, who would impose a tree-like structure on reality, with those who prefer to use the analogy of a web of inter-related aspects of knowledge, as already described. Puzzlingly, the physicist Philip Anderson claims to be a reductionist, at the same time as rejecting the tree description of modern scientific knowledge in favour of the web analogy.[1] Clearly he attaches a different meaning to the word 'reductionism' from that used here—as he has every right to.

There is an immediate and obvious problem with reductionism (in the sense used here). At the present time fundamental physics is not a coherent subject, because quantum theory and general relativity are not consistent with each other. Generations of physicists have struggled with this problem, but the existence of a final 'Theory of Everything' has not been proved. If such a Theory exists, it is *logically* possible that it will include a mental component. However, there is no evidence for this, and reductionists assume that the final Theory will be a recognizable refinement of what we already know about physics. The goal is supposed to be a set of mathematical equations which describe all physical phenomena exactly in a single formalism.

The above view of the future course of physics is described very clearly by Steven Weinberg in *Dreams of a Final Theory*,[2] but it is not shared by all physicists. Indeed, there are many disagreements between those involved in so-called fundamental physics and solid state theorists, not least in relation to the amount of funding each of them receives. In this chapter I argue that the world is too complex and inter-related for an explanation of everything to be possible. (By 'possible' I mean possible in fact, not possible in principle.) The idea of breaking reality up into small parts which are then analysed separately has been brilliantly successful, and may be the only way our type of mind can understand the world. However, its explanatory power is ultimately limited by the existence of chaos and quantum entanglement. There are many complex phenomena which are beyond our understanding, in the sense that even if we knew the relevant physical laws completely, we still could not use them to predict what would happen. If one knows that one cannot make predictions from the relevant laws in certain contexts, one cannot also know that the laws apply in those contexts. While Anderson often expresses himself too strongly, he is, as usual, worth quoting on this matter:

> *Physicists search for their 'theory of everything', acknowledging that it will in effect be a theory of almost nothing, because it would in the end have to leave all of our present theories in place. We already have a perfectly satisfactory 'theory of everything' in the everyday physical world, which only crazies such as those who believe in alien abductions (and perhaps Bas van Fraassen) seriously doubt. The problem is that the detailed consequences of our theories are often extraordinarily hard to work out, or even in principle impossible to work out, so that we have to 'cheat' at various stages and look in the back of the book of Nature for hints about the answer.*[3]

Perhaps the most convincing evidence in support of reductionism is that of chemistry to physics, and we will start by discussing this aspect of the programme. In the 1960s 'real' chemists generally ridiculed quantum chemistry, the discipline which attempted to deduce the properties and reactions of molecules from fundamental quantum mechanical laws. The problem was that the power of computers to perform the necessary mathematical calculations was so limited that only a few very small molecules could be successfully analysed. Since those days matters have changed dramatically, partly as a result of pioneering work by Kohn and Pople, who were rewarded with the 1998 Nobel Prize in Chemistry. The other major factor has been the astonishing increase in the power of computers over the last thirty years. Computations of the shapes and energy levels of molecules of up to a thousand atoms can now be performed routinely. The computed structures are an essential tool in the design of new drugs and the understanding of the dynamics of chemical reactions.

As practised at the present time, the above computations depend upon an ingredient which does not come from quantum theory. Chemists *start* from a knowledge of the approximate geometrical structure of a molecule and *then* confirm and refine this by the use of quantum mechanical laws.[4] This approach is forced on them by the impossibility of carrying out an *ab initio* calculation: for molecules with more than a dozen or so atoms this would be far beyond the resources of any computer which could ever be built. The idea of molecular structure is not easy to justify from first principles because it breaks the symmetries of the quantum mechanical laws. We have already discussed this in the context of chirality on page 198, but related issues arise for other types of isomerism. Reductionists take the view that this is a technical question which does not pose fundamental issues, but there are certainly physicists and chemists who disagree with this assessment.[5]

Let me put it another way. Chemistry involves both knowledge of the laws of quantum theory and decisions to look for molecules of some particular type. Without the latter human ingredient the fullerene molecule drawn on page 189 would never have been discovered, in spite of the fact that it contains only one type of atom, carbon. Nobody could have predicted its existence by solving the Schrödinger equation for a large assembly of carbon atoms from first principles without any preconceptions about the result. The actual process is, *and has to be*, the confirmation using quantum theory of intuitions which come from another source.

One of the most outspoken of the critics of a Theory of Everything is Robert Laughlin, a recent Nobel Prize winner in physics. He and David Pines conclude a deliberately provocative article on this subject with the following words:

> *Rather than a Theory of Everything we appear to face a hierarchy of Theories of Things, each emerging from its parent and evolving into its children as the energy scale is lowered ... The central task of theoretical physics in our time is no longer to write down the ultimate equations but rather to catalogue and understand emergent behavior in its many guises, including potentially life itself ... For better or worse we are now witnessing a transition from*

> *the science of the past, so intimately linked to reductionism, to the study of complex adaptive matter . . .*[6]

The article of Laughlin and Pines has two strands. They believe that understanding the behaviour of assemblies of atoms and electrons is *more important* than chasing 'fundamental' theories, which may well be beyond us. Like Anderson, they also argue that, even within physics, the amount which can actually be predicted has definite limits. In Chapter 6 we examined Newtonian mechanics, one of the most successful of all physical theories, and showed that it contains the proof of its own limitations. The phenomenon of chaos shows not just that there are situations in which we cannot yet predict what will happen using Newton's equations, but that such predictions will never be possible. Weather forecasting has taken this on board by restricting itself to predicting the *probability* that the weather will develop in a certain manner over the following week, and it seems clear that this situation will not change fundamentally. In quantum theory the basic equations are probabilistic, and deny the very possibility of predicting exactly what will happen to an individual quantum particle. It appears that even if reductionism were proved correct, it would be a pyrrhic victory: knowing the equations which govern a phenomenon does not mean that one can thereby know how the phenomenon will develop.

Steven Rose has given a another criticism of reductionism in *Lifelines*. He tells a story in which a physiologist, an ethologist, a developmental biologist, an evolutionist, and a molecular biologist argue about why a frog jumps into a pond. His point is that each explanation is a valuable contribution to understanding the frog's behaviour, and we should not regard any one of them as providing the *real* reason. Rose considers that we should not be trapped into accepting that the most mathematical explanation is also the most fundamental. He regards this view as a consequence of the particular way in which Western science developed. Although a mathematician myself, I entirely agree with him.

Biochemistry and Cell Physiology

The application of the reductionist method to biochemistry has progressed enormously over the last fifty years. The conclusion is inescapable: every individual biochemical reaction in a cell can be explained in purely chemical terms, and hence in terms of quantum theory. Max Perutz has written:

> *Since then [Hopkins'] views have been vindicated by the demonstration that such fundamental and diverse processes as the replication of DNA, the transcription of DNA into RNA, the translation of RNA into protein structure, the transduction of light into chemical energy, respiratory transport and a host of metabolic reactions can all be reproduced in vitro, without even a hint of their individual activities being anything more than the organized sum of the chemical reactions of their parts in the test tube.*[7]

This type of investigation of cell function has been enormously important in extending the limits of our knowledge and giving hope of eventually curing a variety of diseases. There is no need to elaborate since there are newspaper and popular science articles on new discoveries in this field more or less weekly. These provide support for the current commitment of philosophers such as David Papineau to physicalism.[8] This should be contrasted with vitalism, the centuries-old belief that the behaviour of living creatures involves some non-physical vital spirit, which was quite popular during the nineteenth century. Vitalism withered during the twentieth century because no evidence to confirm the existence of such an influence was ever found, in spite of enormous research into physiology.

Even if one accepts physicalism, there is a fundamental difference between cell physiology and the application of Newton's laws to planets. In *Lifelines* Steven Rose has emphasized that the processes going on at different levels in cells are very heavily inter-related: the chains of cause and effect between the various components of a cell go in all directions:

> *Genetic theorists with little biochemical understanding have been profoundly misled by the metaphors that Crick provided in describing DNA (and RNA) as 'self-replicating' molecules or replicators as if they could do it all by themselves. But they aren't, and they can't. . . . particular enzymes are required to unwind the two DNA strands, and others to insert the new nucleotides in place and zip them up. And the whole process requires energy, the expenditure of some of the cell's ubiquitous ATP.*[9]

These are only a few of the activities involving DNA and RNA in the general metabolism of a cell. Instead of referring to DNA as the controller of all higher level processes, it would be just as appropriate to say that the cell uses its DNA to carry out various tasks, such has the manufacture of proteins.

I should add that even a complete account of cell physiology, were that possible, would not enable us to predict the behaviour of an individual cell even a few seconds into the future, except in a laboratory. Let us imagine John, standing on the balcony of a house and looking into the night sky. The passage through the atmosphere of a meteor stimulates his retina and hence the neurons in his cortex. The chemical balance of one such neuron therefore depends upon the strength of the local street lighting, the degree of cloud cover and the trajectories of all near Earth objects. Even if all of the relevant information could be gathered together, it would have nothing to do with John's physiology.

In fact John never sees the meteor. He returns into his house at the critical moment to answer the telephone. The call turns out to be a wrong number. So the state of John's neuron depends on a mistake made by someone else a few seconds earlier, and not upon the meteor after all.

Such examples have prompted Edelman to argue that 'for systems that categorize in the manner that brains do, there is macroscopic indeterminacy'.[10] Even the most committed reductionist has to admit that there is *no way of predicting* many brain events at the purely biological or chemical level. One can only hope to predict the behaviour of individual cells in the tightly constrained setting of a laboratory.

It is legitimate to reply that biochemistry and cell biology are concerned with the general *mechanisms* which govern the behaviour of cells, and not with the accidental circumstances particular cells happen to experience. In this respect the subjects differ from astronomy, mechanics, and engineering, which do predict what will happen to individual bodies very accurately. The reason is clear. Many Newtonian systems are closed to a good approximation, but most biological systems are open, that is heavily influenced by the surrounding environment. One can only claim that biochemistry is reductionist to the extent that it declares non-reducible behaviour to be outside its subject matter. Within these limits it has been extraordinarily successful over the last fifty years. This restriction of the scope of biological science to the elucidation of mechanisms contrasts strongly with the claim of Laplace and others that there is nothing which science cannot explain.

Since I might be misinterpreted, let me make it clear that I am not arguing for the introduction of a new principle to take over where physics or chemistry fail to deliver the goods. What I am saying is that the only explanation that we are ever likely to have for people's behaviour is in terms of motivations, thoughts, preconceptions, love, hatred, etc. The idea that there is a deeper analysis in terms of the motions of the atoms and electric fields in people's brain *and everywhere else in their surroundings* cannot be used to predict the actions of particular individuals. The relevant computations would involve so many internal and external factors that they could not possibly be implemented in practice. The existence of an intermediate level of explanation, involving neural networks and brain biochemistry, may well help us to understand certain types of abnormal brain functions, and even to cure some of them. It is unlikely to change the way in which we describe people's behaviour in everyday life.

Prediction or Explanation

The reader will have noticed that I have frequently referred to the task of science as being to predict what will happen in specified situations. With the above examples in mind one might argue that this is too narrow a view, and that *understanding* is the true goal. Newton provided equations which explain (to a mathematician or physicist) how material bodies move under the influence of gravity. He and later Laplace convinced everybody that the theory was correct by making highly accurate predictions of particular astronomical events. Unfortunately Newton's theory was ultimately superseded by theories whose *explanations* of the same events were totally different even though the *predictions* were almost identical. General relativity and quantum theory use entirely different branches of mathematics from Newton's theory, even though all three yield almost exactly the same predictions of where a stone goes when one throws it. In quantum theory, particularly, the very idea of explanation has undergone a fundamental change. The only fully consistent account of the subject is the set of mathematical equations, and the intuition of physicists is limited to helping them to guess what the equations will predict.

The situation in the biological sciences is quite different. Here prediction of the behaviour of individual entities is regularly impossible. The goal of the subject is rather the elucidation of general mechanisms which control the behaviour of organisms. The sense in which this study is scientific is quite different from that of astronomy. One goal is the possibility of predicting the effects of various types of drugs on the functioning of organisms or cells. A second is the possibility of carrying out surgery on people or animals with the desired effects. Yet another is the possibility of breeding or genetically modifying plants and animals systematically. None of these procedures is ever likely to be one hundred percent reliable, but the increase in our ability to intervene successfully demonstrates that our understanding is genuine. Interestingly the explanations found in this field are of the type which Descartes would have approved of. Biologists study the interactions of material bodies in immediate proximity with each other. Action at a distance is not relevant, and explanations make sense in terms which a layman can often understand, because mathematics plays a much more subordinate role.

Even in solid state physics there is a profound difference between understanding and prediction. Suppose that one had a large computer program which simulated the forces between the atoms of a complex solid, and correctly showed the detailed crystalline microstructures which they can possess. This program would provide no understanding of what was going on. In particular if the parameters of the model were changed slightly, one would have no option but to run the program through again. Understanding in such cases means constructing a mathematical model of the size and type which a human mind can handle, to work out general features of the solid before it is examined.

Even if we agree that understanding the natural world is the true goal of scientific activity, its validity can only be demonstrated if the theories which we devise can be tested. A theory may be confirmed in a wide variety of ways, and we must be very careful not to take any one science as the model for all others in this respect. The progress of physics shows that understanding is an elusive matter, and that no matter how well a theory performs, it may be superseded by a quite different one in the future. As society develops, different ways of understanding will appear and compete with each other on the basis of their simplicity and scope. We should not be too confident that a single method of understanding all phenomena will ultimately emerge—the world is far too complex and our brains far too limited for this to be likely.

I have avoided one issue in order to avoid being sucked into a deep philosophical problem: giving a general definition of understanding. This is harder that it appears, and many different solutions have been proposed.[11] Rather than discuss these, let us consider an example which illustrates the difficulties. Any historian of science could provide many similar cases.

When Newton proposed his law of gravitation, he was acutely aware of not having explained how two distant objects could exert a gravitational force on each other. In the seventeenth century such a possibility was simply not acceptable to many people, and he was not able to reply effectively to his critics. By 1760, however, his theory was completely accepted, and nobody regarded this as a problem. No explanation for action at a distance was considered necessary: this was simply how nature worked. When Einstein's general theory of relativity appeared the situation changed again. Action at a distance was again barred, and even more emphatically; in addition the gravitational force disappeared, to be replaced by a varying curvature of space-time. It therefore appears not only that explanations change with time, but even that *what needs explanation* changes. What is acceptable as a theory depends upon the social context of the time, however unpalatable that may be to those who would like to banish human beings from the science in which they engage.

Money

In the previous sections we have considered some successes and failures of the reductionist programme in chemistry and physics. We turn next to a subject discussed by Donald Gillies[12] in which it is difficult to argue that any reductionist account can be given—money. Let us start with a potted history.

Until the twentieth century money could be equated with pieces of copper, silver, or gold. For most of the time since their introduction in the first millennium BC coins were regarded as intrinsically valuable, their value lying in the metal of which they were made. The introduction of milled edges on coins in the late seventeenth century was intended to stop people clipping the edges, a practice which would make no sense today. At that time and until quite recently paper notes were contracts: British ten pound notes still bear the words:

I promise to pay the bearer on demand the sum of ten pounds

with the confirming signature of the chief cashier of the Bank of England. This promise is literally meaningless! During this classical period coins were real money, while notes were substitutes introduced for the sake of convenience.

The abandonment of the Gold Standard, led by the United Kingdom in 1931, was an acknowledgement of a new monetary theory in which coins became merely tokens, rather than the real thing. In the latter part of the twentieth century accounts in banks became increasingly computerized, and cash-based transactions became ever less important. Money now is an agreement to give that name to certain data stored in machines in banks, and coins form only a small part of an abstract conceptual system.

We thus see that at various times in history money has consisted of lumps of metal, pieces of paper, and more recently magnetic domains on the hard disks of computers. The only thing which has remained constant is society's requirement, enforced by law, that people should honour obligations recorded in these various ways. Thus money should be considered as a *social construction*.

Both its existence and its nature depend upon collective social agreements. Although it affects the behaviour of individuals, it is not merely a part of their mental worlds. If a single individual believed that he/she possessed some money this would not have the desired consequences unless others agreed with this belief. One cannot deny that money is real, because it has real effects upon the physical world. On the other hand this does not give it a Platonic status, because it is not eternal. It did not always exist and its nature may change with time.

The above discussion of money illustrates a general point. One could provide similar arguments for other social constructs, such as the legal system or the French language. These are undoubtedly real if reality is proved by having an effect on material objects. They are also obviously not material objects themselves, and depend for their existence on society. All such examples undermine the reductionist position. They provide examples in which the behaviour of material objects is explained in terms of something higher in the list which we presented at the start of this chapter.

Information and Complexity

A nice example of the futility of looking for reductionist accounts of ecological systems has been given by Alan Garfinkel.[13] Consider populations of rabbits and foxes, in which the foxes eat the rabbits, and both reproduce and die. At the simplest level this may be described by just two variables, the numbers of foxes and of rabbits, together with a linked pair of equations which describe how these numbers change with time. An excessive number of foxes makes the number of rabbits plummet catastrophically, after which the foxes start starving to death. A very small number of foxes allows the rabbits to breed freely, with the result that their numbers explode. The questions to be resolved are whether one should expect the numbers of foxes and rabbit to settle down to an equilibrium, or whether the sizes of the populations will oscillate periodically in time.

We are not concerned here with the details of the mathematics, but rather with what is *not* included in the above description. A reductionist would have to claim that the equations are approximations to a fuller description which involves individual rabbits being eaten by individual foxes. Now such an event is not an abstract one. It has to occur at a particular place and time. Whether a fox eats this rabbit rather than that one depends on where the rabbits are in relation to the fox and how alert to the presence of foxes they are. It also has to bring in factors such as the lighting, amount of undergrowth and whether or not it is raining. A full reductionist explanation would involve so many such factors and be so complicated that the relevant data could never be collected.

This is not to say that somewhat more complicated and realistic models of the rabbit-fox populations cannot be constructed. The point, rather, is that there is a trade-off between how many factors the equations take onto account and whether the model is in practice soluble. Selecting the relevant factors cannot be done from first principles. Ecologists rely upon their experience and judgement to devise models which are simple enough to be solved but complicated enough to capture the essential features of the ecosystem. In fields which such a high degree of complexity, a reductionist approach has no chance of being implemented.

Some scientists working on the theory of complex systems adopt an *anti-reductionist* approach to understanding. They claim that in many cases the use of reductionist techniques would actually be a *barrier* to understanding. If quite different systems exhibit the same detailed behaviour, then the key to understanding depends upon information theory, a relatively new subject. The following examples illustrate what is at issue.

The first is the theory of fluids. It is believed that the behaviour of fluids is governed by the Navier–Stokes equations, and enormous efforts have been put into finding approximate solutions of these under a variety of initial conditions. There have also been derivations at varying levels of rigour from the underlying atomic dynamics. Now consider the fact that at room temperature and pressure water, olive oil, mercury, and butane are all liquids. These substances have totally different molecular structures, but this fact is not relevant to the fact that they all satisfy the Navier–Stokes equation. In most ordinary situations the details of the molecular dynamics is precisely what we do *not* wish to consider when we study the behaviour of these liquids.

A similar point has been made in the study of the mind by the function-alist group of philosophers. They emphasize that the brains of monkeys and octopuses have totally different neural architectures, and evolved independently, and yet they are both conscious of their environments. If we ever meet intelligent aliens we can be sure that their brains will have a very different structure from our own. Functionalists claim that we can only understand consciousness fully if we concentrate on the way in which brains process information rather than on their particular anatomy.[14]

The mathematical theory of games was developed by von Neumann and Morgenstern in 1944 to provide a quantitative basis for studying competition in economics. The subject investigates optimal strategies, that is the rules which players should follow in order to maximize their gains or minimize their losses, under the assumption that the other players are also doing the best possible for themselves. It has had a wide variety of applications, including predicting whether animals should be aggressive or submissive when competing for mates, and how to minimize the risk of losing a nuclear war. (It was widely believed in the 1950s that such a war would have winners and losers.) The theory of such games defies any description in the categories associated with reductionism. The physical composition of the players is wholly irrelevant, while their goals, which belong to the top level of the hierarchy, determine how they should play. Nevertheless the theory does govern how people and animals behave in the relevant situations.

The new and vigorous science of complexity theory (or more precisely self-organized criticality) started in 1987, when the physicists Bak, Tang, and Weisenfeld simulated the growth of a pile of sand when grains slowly trickled onto it. They confirmed that the sand forms a roughly conical shape with a characteristic angle. Their numerical calculations showed that piles had periods of stability, interrupted by sudden landslides. There were landslides of all sizes, the proportions of different sizes being related by universal scaling laws. They argued that the behaviour which they observed might be a *universal* structural feature of complex systems. In this sense the subject is anti-reductionist: it is a study of generic features and not of the consequences of particular physical laws.

Subsequently it has been found (or perhaps claimed) that such considerations apply to the frequency and size of forest fires, earthquakes, stock market crashes, the extinction of species, wars, and many other subjects in which there are sudden catastrophes. However, this is a very young subject, and *actual* experiments on granular piles, with rice instead of sand, have shown that self-organized criticality is *not* a universal phenomenon. It appears to depend upon the shape of the grains and whether the predominant motion of a grain is sliding or rolling.[15]

There are two morals to be drawn from this story. The first is to be cautious about any new scientific discoveries. It is important to wait for the considered judgement of the community, which may take years to emerge. The second moral is that in those cases in which self-organized criticality does occur, it is pointless to try to prevent individual catastrophes, because this will merely postpone the day of reckoning. One can solve the problem, but only by changing the behaviour of the system in a more fundamental manner.

Subjective Consciousness

We now pass to another subject in which there has been prolonged and heated debate about the relevance of the reductionist viewpoint. This is the study of subjective consciousness (SC). This is what we experience ourselves, to be contrasted with third person consciousness, which is what we can infer about others by observing their behaviour. Its reality seems undeniable, even though it is very difficult to discuss its nature. All attempts seem to move more or less rapidly to the third person subject. Thomas Nagel even states that the problem of consciousness is rendered insoluble by the very assumptions of the Cartesian philosophy of science:

> So when science turns to the effort to explain the subjective quality of experience, there is no further place for these features to escape to. And since the traditional, enormously successful method of modern physical understanding cannot be extended to this aspect of the world, that form of understanding has built into it a guarantee of its own essential incompleteness—its intrinsic incapacity to account for everything.
>
> One consequence is that the traditional form of scientific explanation, reduction of familiar substances and processes to their more basic and in general imperceptibly small component parts, is not available as a solution to the analysis of mind. Reductionism within the objective domain is essentially simple to

> *understand ... No correspondingly straightforward psychophysical reduction is imaginable, because it would not have the simple character of a relation between one objective level of description and another.*[16]

Let us start with the perception of pain. At first sight it appears that a reductionist account of pain should be straightforward. There are specific receptors in the skin and elsewhere which respond to certain stimuli, such as extreme heat, cold, certain chemicals, and physical damage, and when these receptors fire we often feel pain. Detailed descriptions of the structure of the relevant receptors and of the mechanisms which activate them are readily available. Unfortunately this does not settle the matter. Under the influence of anaesthetics we may not experience the pain which we normally would. At the other extreme the phenomenon of phantom pain from limbs which have been amputated long ago is well attested. Again, sports players frequently continue playing unaware that they have suffered quite severe injuries because of the excitement of the activity. Thus we discover that pain sensations are by no means simply and directly related to messages originating in the periphery of the nervous system. The next task is to determine what physical events within the brain correspond to the subjective experience of pain. If we believe that octopuses experience pain, then a complete account of pain cannot be dependent upon the particular brain structures which mammals possess. Making progress on this is going to be a very difficult task.

Unfortunately the above fails to address the nature of the subjective experience of pain. If we experience pain *exactly* when some brain mechanism is in one particular state, is there a difference between the experience of the pain and the physical operation of this mechanism? We will only mention the following three positions, each of which has many subvarieties:

- Epiphenomenalism. Subjective experiences are distinct from the operation of brain mechanisms. They accompany it but have no actual effect upon the way in which the brain operates.
- Interactionism. The relevant brain mechanisms do not obey the normal laws of physics, but can be affected by subjective experiences in a way which transcends physics.
- Physicalism. Subjective experiences are completely explained by the operation of the relevant brain mechanisms, which are determined by the laws of physics.

It is well recognized that each of these positions leads to awkward problems if pursued sufficiently far. Some of these will be described in the next few sections.

The Chinese Room

John Searle has had a long interest in the philosophical problem of distinguishing between simulation of consciousness and the real thing. He argued that even if a computer could simulate a human conversation perfectly, it would no more be conscious than a computer predicting the weather has an actual storm inside it. To explain the distinction he invented a Chinese room story which runs as follows:

> *Suppose a computer program can be designed which will accept questions in Chinese and process these by a complicated set of rules leading to a response, also in Chinese. If the responses are always appropriate, does that mean that the computer is conscious and understands Chinese?*[17]

To prove that the answer is no, Searle imagines a person who does not understand Chinese and who is put in a room with the complete manual of instructions which the computer would follow. On being set the question he eventually works through all the instructions and gives the response without having any idea of their significance. Where is the consciousness, or more specifically the understanding of Chinese, in this case, he asks? He claims that a computer translation program must act syntactically and does not touch on the semantics, i.e. the meaning, of sentences. The difference is illustrated by the following sentences:

> *I was surprised to see that Joan was wearing a red dress.*
>
> *I was surprised to see that John was wearing a red dress.*

A translation program would deal with the two sentences more or less identically, possibly changing the form of the second verb in an inflected language. But we understand that surprise means something and look for that meaning. Because we know something about social contexts, we guess in the first case that Joan was thought not to like the colour red, but in the second case that John does not usually wear women's clothing. (Both guesses might of course be wrong.) Without the social context, which is not needed for the translation, we would not be able to hazard any guess about the significance of the sentence.

Now change the word 'surprised' to 'pleased'. The first sentence suggests that I have a warm relationship with Joan, and that I think that the colour red suits her. The second is rather strange, and impossible to interpret without further context. But from the point of view of a translation program almost nothing has changed.

Over a hundred articles relating to Searle's argument have been published, and it seems safe to say that the resulting disagreements can no longer be resolved. The extent to which his example is relevant to the existence of SC is also debatable. I will raise only one of the most common objections to Searle's argument, without any suggestion that it is the most important one.

A well known response is that the understanding of Chinese belongs to the whole system rather than to the parts. In the same way individual neurons in a person's brain do not understand a message, and the understanding is a result of all of the neurons working together. A vivid way of expressing this is to point out that if one looks at the engine of a car, one cannot identify a particular part which makes it work. Its 'engineness' is a consequence of all of the parts working harmoniously together. It is implied by some that Searle's argument is analogous to vitalism, the outmoded philosophy that inanimate, even organic, matter could not gain life without the addition of a 'vital spirit'.

This has been repudiated as irrelevant, on the grounds that consciousness is both a property attributed to a person by others on the basis of behaviour, *and also* something experienced as a subjective fact. This second aspect has no analogue for engines. Even if we do not know the biological mechanism which causes SC, that does not allow us to pretend that it does not exist. We are conscious of certain brain functions but not of others, and this *must* be explained. Consciousness is one of the primary facts about ourselves to be explained, and any theory of the brain which does not do this is necessarily incomplete.

Searle accepts that in principle an artificial system could be conscious, but only if it had the appropriate (and so far unknown) internal structure. Like many other philosophers, he rejects the Turing test for consciousness. This asks whether a computer can converse (perhaps by email) with a person in such a way that the person cannot tell that it is 'only' a computer. Perhaps, however, the debate about the Turing test is predicated on a false hypothesis. The failure over several decades of attempts to get computers to simulate human behaviour in open ended contexts suggests that this might only be possible if the computer does indeed contain the appropriate semantic structures. Expert systems such as chess playing programs do not, and they only work in very narrowly constrained contexts.

Zombies and Related Issues

In his recent book *The Conscious Mind* David Chalmers dismisses physicalism, stating that the existence of subjective consciousness is not capable of serious doubt—even though he admits that Daniel Dennett and others do doubt it. He agrees that denying the existence of SC makes explaining the world much simpler, but reiterates that one must take SC seriously. We will discuss physicalism further in the next section.

One of Chalmers' arguments against interactionism imagines that there exist 'psychons' in the nonphysical mind which may affect physical processes in the brain and which are themselves the seat of subjective experience:

> *We can tell a story about the causal relations between psychons and physical processes, and a story about the causal dynamics among psychons, without ever invoking the fact that psychons have phenomenal properties. Just as with physical processes, we can imagine subtracting the* phenomenal *properties of psychons, yielding a situation in which the causal dynamics are isomorphic. It follows that the fact that psychons are the seat of experience plays no essential role in a causal explanation, and that even in this picture experience is explanatorily irrelevant.*[18]

Other descriptions of how subjective consciousness might act on the brain seem to have a similar problem. They seem attractive at first sight because they correspond to the way we feel inside ourselves. However, once one starts to analyse the idea in detail it seems impossible to build subjective experience into any *systematic account* of its mode of action on the brain. The very process of providing an explanation seems to transfer the essence of the phenomenon to a different place.

Much of Chalmers' book explores a type of epiphenomenalism. He spends a considerable number of pages discussing zombies, which have *exactly* the same physical structures in their brains and behave in *exactly* the same way as we do, but which do not have any inner subjective experiences to match their behaviour. He puts forward plausible arguments designed to show that zombies do not exist and that all objects sufficiently like ourselves must in fact possess SC. These arguments are based upon imaginary scenarios of the type which philosophers delight in considering. The following is not in his book, but has a similar flavour. Let us suppose that a half of all humans are zombies, and the rest, including the reader of course, possess SC. If asked, the zombies would state that they possessed SC, and would appear to be indignant at any suggestion that they did not, because they behave exactly like us. They would be wrong, but would not know this because, as zombies, they can know nothing in any true sense. Now suppose that a zombie and one of us marries and has children. Perhaps the children would all be zombies, perhaps they would all possess SC and perhaps this would be a matter of chance. Alternatively our distant descendents might possess a fraction of our SC depending on the proportion of ancestors who were zombies. The problem here is that it is difficult to imagine what having a fraction of 'normal' SC might mean.

Let us turn to the much-discussed subjective experiences of colours. These are frequently called qualia to distinguish them from psychological processes which can be studied by scientific methods. Consider the possibility that A may look at a red object and call it red, because he has been taught to do so, even though his subjective experience is what B would have when B sees a green object. This philosophical possibility of subjective colour inversion was first discussed by John Locke. He already stated that there is no way in which this phenomenon could be detected, since the subjective impressions of an individual are not available for external inspection. All one can do is compare people's responses to similar objects. In such discussions it always seems to be assumed that this is merely a 'philosophical' possibility, and that we have some other reason to believe that different people looking at a red object have the same subjective experience, unless one of them has a visual defect. Actually the detailed neural connections in any two individuals are so different that their subjective experiences might well differ as much as, say, their ability at mathematics or portrait painting. We know that trained musicians can distinguish the separate instruments in an orchestra in a way which others simply cannot. Their training has altered their brain circuits, and hence their subjective impressions.

It appears that even if one believes that SC is a real phenomenon, we have no way of describing it publicly. This being the case, it might be as well for us to give up trying. Chalmers' attempt to develop a science of SC is bound to fail because ultimately it is only based on plausibility arguments. His thought experiments cannot be a substitute for real-world experiments, even though they are interesting from a philosophical point of view. They may enable one to rule out certain types of explanation of SC, but in science it has repeatedly been found that *real experiments* are needed to obtain a correct description of the world. Unless some radically new ideas appear in the future, it appears that our belief that others possess SC similar to our own is not amenable to rational or scientific proof. SC cannot be studied, only experienced. Third person consciousness, on the other hand, can be investigated by a rapidly increasing range of techniques.

A Physicalist View

Inevitably there are philosophers who deny the existence of SC. A recent collection of essays by Churchland and Churchland, *On the Contrary*, argues that our current views about consciousness are a part of a profoundly mistaken folk psychology (FP):

> *FP functions best for normal, adult, language-using humans in mundane situations. Its explanatory and predictive performance for prelinguistic children and animals is decidedly poorer. And its performance for brain-damaged, demented, drugged, depressed, manic, schizophrenic, or profoundly stressed humans is pathetic. Many attempts have been made to extend FP into these domains, Freud's attempt is perhaps the most famous. All have been conspicuous failures.*[19]

The Churchlands describe in some detail current research on the brain, considered as a self-programming neural network. This constantly and sometimes incorrectly tries to find the best mental pattern from a vast space of possibilities to match the images, ideas, etc. being considered. This idea fits well what we learnt about the functioning of our visual systems in Chapter 1. The absence at the deepest level in our brains of a sentence-like or propositional structure explains the failure of AI systems to reproduce anything like human cognition. They argue that since FP is based on introspection it fails to appreciate that concepts such as subjective sensations which appear to us to be unitary and irreducible may actually be highly complex. They also anticipate a major change in the way we describe our subjective thought processes as a result of current scientific advances. In this context their dismissal of the arguments of Nagel and Searle above is hardly surprising. Consider the following passage:

> *There is also a standard and quite devastating reply to this sort of argument, a reply which has been in the undergraduate textbooks for a decade . . . Stated carefully the argument has the following form:*
>
> 1. *John's mental states are known-uniquely-to-John-by-introspection.*
> 2. *John's physical brain states are not known-uniquely-to-John-by-introspection. Therefore, since they have divergent properties.*

> 3. *John's mental states cannot be identical with any of John's physical brain states.*

> *Once put in this form, however, the argument is instantly recognizable to any logician as committing a familiar form of fallacy, a fallacy instanced more clearly in the following (example).*

> 1. *Aspirin is known-to-John-as-a-pain-reliever.*
> 2. *Acetylsalicylic acid is not known-to-John-as-a-pain-reliever.*
>   *Therefore, since they have divergent properties.*
> 3. *Aspirin cannot be identical with acetylsalicylic acid.*[20]

The Churchlands point out that in the second case the paradox disappears as soon as John is told that acetylsalicylic acid is identical with aspirin. Such arguments are said to be epistemological, that is about people's knowledge. However, Searle claims that his argument is quite different: it is intended to demonstrate a true difference between first person and third person consciousness. In other words it is an ontological argument, that is about the nature of things. In response to this the Churchlands state that Searle's conclusions do not follow from his premises. They are simply assumed from the start.

Let us put this technical debate aside, and return to the central issue. Suppose that circuits are discovered within the human brain such that the activation of these circuits occurs *precisely* when the subject claims to experience the sensation of pain. Suppose also that the linkages between these circuits and the pain receptors in the peripheral nervous system are understood, and the mechanism by which local anaesthetics act to stop the activation of these circuits is also discovered. Are we really expected to believe that these facts would have no implications at all for the explanation of the subjective experience of pain? At the very least one should admit that the problem of pain would look very different after such scientific discoveries. If a similar reduction is eventually made for all subjective experiences it is possible that interest in SC will go the same way as vitalism did.

Notes and References

[1] Anderson 2001

[2] Weinberg 1993

[3] Anderson 2001

[4] I am referring to the fact that in the Born–Oppenheimer approximation, the configuration of the nuclei is chosen rather than computed from first principles.

[5] See Hendry 1998, Hendry 1999, Laughlin and Pines 2000 and references there.

[6] Laughlin and Pines 2000

[7] Perutz 1989, p. 219

[8] Papineau 1990, Papineau 1991, Papineau 2001

[9] Rose 1997, p. 127

[10] Edelman 1995, p. 203, 204

[11] See de Regt and Dieks 2002 for a recent discussion which follows a similar line.

[12] Gillies 1990

[13] Garfinkel 1991

[14] Warner and Szuba 1994

[15] Frette et al. 1996

[16] Nagel 1994, p. 66

[17] Searle 1980, Searle 1984

[18] Chalmers 1996, p. 158

[19] Churchland and Churchland 1998, p. 32

[20] Churchland and Churchland 1998, p. 117

10
Some Final Thoughts

This final chapter is a collection of topics which relate to the status of science in our society, and to its conflicts with other systems of thought. The matters discussed include the so-called 'anthropic principle' and cultural relativism. In some cases I point out definite errors in standard arguments. In others the plausibility of the criticisms of science depends upon beliefs which are imported from outside science. There is nothing wrong with doing this, provided one is honest about it. Science is a system of thought, and should not claim to have a monopoly on the truth. But neither can its achievements be dismissed as of no import. I finish with a statement of my overall conclusions.

Order and Chaos

The science of thermodynamics grew during the nineteenth century as a consequence of attempts to make steam engines more efficient and to find the ultimate limits on what was possible in this new technology. During that period many people tried to design perpetual motion machines, some of extraordinary ingenuity, but without exception they failed to work. Their failure was encapsulated in the first law of thermodynamics, also called the law of conservation of energy. It states that you cannot get something for nothing, or that perpetual motion machines are impossible. The evidence for it is so convincing that applications for patents for such machines are now rejected without consideration, in Britain and the USA at least.

The ideas behind the first law are not elementary. Kinetic and potential energy were both well known in Newtonian mechanics, but for the purposes of the first law one also has to include heat as a form of energy. When a ball falls to Earth and eventually stops bouncing it loses its potential energy (which it had by virtue of its initial height), and also its kinetic energy (which it had just before hitting the ground by virtue of its speed of motion) with nothing obvious to show for these. During a bounce its energy is converted to elastic energy of the material of the ball, before being reconverted into kinetic energy as the ball moves upwards. The process is not entirely efficient and on each bounce a little of the energy is lost within the material of the ball as heat. Eventually the mechanical energy is converted entirely into heat. This is not just a manner of speaking. One can quantify heat energy using the temperature of the ball and its heat capacity, and confirm that the total energy is indeed conserved.

The second law of thermodynamics is regularly misinterpreted by creationists. One formulation is that there is a fundamental irreversibility in the universe, in which energy is converted into more degraded forms, and a quantity called entropy inevitably increases. It may be summarized as asserting that the overall disorder of a closed system always increases. An example of this occurs if one pours a cup of hot water and a cup of cold water into a jug. The two mix together producing a jug of water at an intermediate temperature which can be calculated using the first law. The second law implies that the water in the jug will never unmix itself into two halves at different temperatures. It also ensures that a ball resting on the ground will never start shaking more and more vigorously until it bounces into the air, unless there is some external reason for this (e.g. an earthquake).

Unfortunately this law has frequently been misinterpreted as saying that order can never emerge from chaos, leading to the claim that the existence of living creatures is proof of the existence of an creator. This argument is simply wrong. The second law refers to the behaviour of a *closed* system, that is one which is not interacting with the external world, and which is therefore moving steadily towards thermodynamic equilibrium. However, most of the phenomena in which we are interested concern *open* systems, far from equilibrium. The dynamics of most life on Earth is entirely dependent upon the constant flow of heat and light from the Sun. Only when the Sun runs out of energy and the core of the Earth cools down, billions of years in the future, will the Earth be an isolated equilibrium system, and will life cease to exist. In the meantime complex structures are driven into existence by the flow of energy from the Sun to the Earth, and then from the Earth to outer space.

Everybody has probably seen a ball balanced on the top of a jet of water. It is surprising that it does not immediately fall sideways out of the jet and then drop to the ground. But it is quite stable: it does not need to be placed in position with extreme care, but moves back to the centre of the jet if slightly displaced. This is a typical example of a system which can stay in a state far from its natural equilibrium (lying on the ground) as long as there is a constant flow of energy (moving water) to sustain it.

The possibility of complex patterns emerging by purely physical processes from materials which contain no traces of the patterns is beautifully illustrated in the structure of snowflakes. Figure 10.1 is just one out of an astonishing 2453 different examples in *Snow Crystals*[1] by Bentley and Humphreys. William Bentley produced the first ever photograph of a snow crystal in 1885 and discovered that no two were identical. Almost all of the images in the book have approximate hexagonal symmetry.

Snowflakes arise by the accumulation of water molecules which condense as ice onto a central nucleus. Their symmetry is a result of the atomic interactions between the constituent particles, while the slight variations between different

Fig. 10.1 A Snowflake
Reproduced from the W.A. Bentley collection by kind permission of the Jericho
Historical Society

snowflakes depend upon the varying temperature and humidity of the cloud
within which they form and through which they fall. Nowhere is a designer
needed, in spite of the astonishing beauty and regularity of the snowflakes. The
creation of snowflakes does not contradict the second law of thermodynamics,
since the conditions within the cloud guarantee that the local environment of
each snowflake is sufficiently far from equilibrium. Their symmetry is a consequence of the atomic forces between the water molecules which form them,
and the fact that at any instant the environment of a growing snowflake is the
same on all sides of it.

Of course snowflakes disappear as fast as they grow, so comparing them
with living organisms is only an analogy. Nevertheless, the fact that such highly
organized structures can indeed appear out of nothing within a few minutes
proves that order can appear 'from nowhere', and also makes it more plausible
that given billions of years vastly more complicated entities such as living cells
might appear by purely physical processes.

As a second example consider soap bubbles. The shape of such a bubble,
almost a perfect sphere, is not designed by the person blowing the bubble, nor is
the shape stored in the soap mixture waiting to be released. Except for the fact
that our reactions are dulled by familiarity, it is highly surprising that blowing
on a bit of liquid can have such a result. Once again no designer is needed.
As soon as it is formed the bubble changes shape to minimize its energy in

conformity with the thermodynamic laws, and it can be proved mathematically that the minimum energy configuration is a perfect sphere. The same applies to mud bubbles in natural cauldrons.

The idea that order may emerge from chaos by the operation of purely physical laws was pressed by Ilya Prigogine, who received a Nobel Prize in 1977 for his development of this subject. He identified many situations in physical chemistry and physics in which the behaviour of the associated nonlinear systems gave rise to organized behaviour. These include the spontaneous formation of Bénard cells (periodic structure in space) in fluid convection and the variety of oscillating chemical reactions (periodic structure in time) going under the name of the Belousov–Zhabotinsky reaction.

The examples above do not *prove* that highly organized structures such as living cells *must* have emerged by natural processes. They were only intended to counter the argument that this is physically impossible. There is no knock-down argument to determine whether the very first life was created or came into existence by natural processes. The events concerned are sufficiently remote that settling the issue is going to be very hard.

Anthropic Principles

The progress of science has been accompanied by separating the question of *why* the laws of nature are as they are, from the detailed examination of the laws themselves. A few theoretical physicists turn to the big questions, and discuss what they call anthropic principles. Among these one has to mention John Barrow, Paul Davies,[2] John Polkinghorne, and Martin Rees.

The debate centres around the fact that the laws of nature appear to be very finely tuned by the values of certain fundamental constants involving the weak and strong interactions. In the early 1950s Fred Hoyle discovered that the production of carbon in stars, and hence the appearance of life as we know it, depended on the existence of a resonance in the carbon nucleus at a certain energy and hence on the precise values of the fundamental constants in nuclear physics. This was only the first of a number of discoveries that small changes in the fundamental constants of physics would have a profound effect on the evolution of the universe. Almost any such change appears to prevent the complex structures and thermodynamic disequilibrium on which our existence depends. These facts led Brandon Carter in the early 1970s to formulate the weak anthropic principle—that the existence of human life imposes certain conditions on the universe, since its structure must be consistent with our being here to observe it.

Attitudes towards this principle may be classified as follows—the phenomenon is real and implies the existence of God; there exist many different universes, each with its own values of the fundamental constants; the whole debate is overblown and unscientific. This summary is, of course, too simple-minded, but it will serve to set the scene.[3]

Let us start with the first response. John Polkinghorne has written:

In the fine-tuning of physical law, which has made the evolution of conscious beings possible, we see a valuable, if indirect, hint from science that there is a divine meaning and purpose behind cosmic history.[4]

He goes on to say that the evolution of conscious life seems the most significant thing that has happened in cosmic history, and we are right to be intrigued by the fact that so special a universe is required for its possibility. This statement is difficult to disagree with, since any entity capable of doing so is presumably conscious. (Douglas Adams's android Marvin would presumably disagree, however!) On the other hand Polkinghorne does not claim that the anthropic principle provides a *proof* of the existence of God, the so-called argument of design, and theologians in general are rather careful not to over-interpret the scientific facts. In *Universes*[5] John Leslie, on the other hand, argues that such caution is inappropriate and that the evidence for design is overwhelming.

At the other extreme are physicists such as Heinz Pagels, who believes that the influence of the anthropic principle on the development of contemporary cosmological models has been sterile: it has explained nothing, and has even had a negative influence. When reviewing Stephen Hawking's book *The Universe in a Nutshell*, Joseph Silk referred to the anthropic principle as 'one of the more remarkable swindles in physics'.

There are in fact a few possible lines of investigation of the principle which hold out some slight possibilities of being scientifically testable. Martin Rees and others have discussed the possibility that inflationary cosmological models and other more speculative theories might allow the existence of myriads of different universes in which the fundamental constants have different values.[6] If this is correct then the values of the constants in our particular universe, which would be just a part of a much vaster 'multiverse', must allow us to have evolved. No theological conclusions need be drawn from the coincidences. It would be relevant to know roughly how tiny our part is within the greater whole.

The very existence of the numerical coincidences underlying this debate has recently been questioned by Robert Klee,[7] in an article provocatively entitled 'The revenge of Pythagoras: how a mathematical sharp practice undermines the contemporary design argument in astrophysical cosmology'. He likens the search for numerical coincidences in the fundamental constants to the mystical numerology of the Pythagoreans, and provides detailed evidence that the coincidences are much less impressive than is usually claimed. This paper dissects the scientific literature on the problem, and should be read by anyone with a serious interest in it.

Bogus coincidences are extremely easy to produce. Consider the following:

$$\frac{m_p}{m_e} = 1836.153$$

$$6\pi^5 = 1836.118$$

where m_p is the mass of a proton and m_e is the mass of an electron. It looks impressive, and is far better than many of the coincidences 'noticed' by those arguing for design. If pressed one might even 'justify' the power 5 as being one half of the dimension of the currently popular model of string theory. But I found it simply by playing around for a few minutes with powers of π. Although scientists noticing unexpected numerical coincidences in particle physics and cosmology are not consciously doing this, it may be the true explanation of what they have found.

This is, of course, very unsympathetic. The occurrence of the same constant in quite different contexts often leads scientists to discover important new connections between different phenomena. This has been particularly true in the study of critical exponents in statistical mechanics. It is a mistake to generalize about such issues, but Klee's conclusion is that the cosmological evidence for design is far from compelling. Nor is he the only one. Livio and colleagues constructed a detailed computer model of stellar interiors in order to find out the effects of slightly changing the carbon resonance mentioned above. They concluded by stating, with typical academic caution, 'we believe that at least some formulations of the strong anthropic principle (are) weakened significantly by our results'.[8] These disagreements between experts about whether there is anything to be explained are unsettling, to say the least.

Let us look at the issue from a different perspective. There has been a long history of attempts to invoke the hand of God to explain matters which science currently could not. As science developed, this led to a series of tactical withdrawals by theologians, and the whole idea of invoking a 'God of the gaps' has been discredited by many theologians themselves. Yet the anthropic principle is of precisely this type: it depends upon the view that the fundamental constants could have taken any other values and that a substantially less arbitrary model of the universe will never be found. Sixty years ago it would not have been possible to formulate the principle, because the evolution of the stars was not sufficiently well understood. Over the last few decades the main task of theoretical physicists has been to go beyond the standard model. Their goal is to find a theory which would enable the number of fundamental constants to be reduced from eighteen, possibly to none in the hypothetical Theory of Everything. Deliberately or not, believers in the anthropic principle are encouraging the view that there is no scientific way of explaining why the constants have the particular value which they do. They may be right, but the best way of finding out is to try to find a reason.

Let us next concede the possibility that the universe might indeed have been designed for the production of carbon as Hoyle and others have suggested. It is plausible that there are many planets (billions) on which carbon-based life has developed, since the evidence from our own planet is that life appeared almost as soon as the conditions made it possible. Granted this, we can deduce almost nothing about the nature of the designer(s). It might be a super-civilization, of the type favoured by many science fiction authors. It might be an entity which created the universe out of mere curiosity, as the mathematician John Conway did with the computer Game of Life. In spite of the arguments of Leslie, I see no reason to assume that the creator has any *ethical or religious* purposes in producing the universe. Perhaps he is a super-chemist, and is just using organisms to manufacture a variety of proteins with as little effort as possible. Even if the creator was interested mainly in the possible evolution of life, there is no reason to believe that life on the planet Earth was uppermost in his mind. Perhaps he was interested in a much more ethically responsible species which will evolve far away in the universe a billion years in the future.

One driving force behind people's search for proofs of the existence of God is the need to give ourselves a special status in the universe. We crave the security and meaning that a benevolent creator—a personal God—would provide in our lives, and must be careful not to let our wishes influence our judgement. There is, unfortunately, plenty of evidence of the contingency of our individual lives, including the fact that each one of us only exists because of the success of a particular chain of acts of copulation stretching over many millennia. We know that the Earth has been hit by massive asteroids on several occasions, and that there have been many super-volcano eruptions, some quite recently in the past. Both types of event have had disastrous effects on the ecosystem, and will no doubt cause huge numbers of deaths in the future. The author finds it hard to reconcile such events with belief in a benevolent God, although he would like to be able to do so.

From Hume to Popper

Between the seventeenth and twentieth centuries the dominant philosophy of scientific discovery was very straightforward. Scientists amassed a large number of observations, and then searched for a simple explanation, possibly by a set of mathematical equations. The process was supposed to be objective and the laws obtained were believed to be *true*, subject to the normal provisos about possible errors. Doubts about the justification for believing in the truth of scientific laws might be expressed by philosophers such as Hume, but most scientists were confident that these doubts need not cause them any loss of sleep. During the first thirty years of the twentieth century they were to discover that much of their prized knowledge of the world needed radical revision. Many facts, previously certain, turned out to be no more than approximations to a quite different truth. This led to a re-examination of the basis for the possession of any final knowledge of the world.

We take as the starting point for this story the radical scepticism of the eighteenth century philosopher, John Hume, as described in his *Treatise of Human Nature*. Hume called into question the basis for the acquisition of any knowledge about the outside world. He emphasized that the repeated occurrence of an event B after an event A does not *logically imply* that A causes B:

> *Let men once be fully persuaded of these two principles, That there is nothing in any object, considered in itself, which can afford us a reason for drawing*

*a conclusion beyond it; and That even after the observation of the constant
conjunction of objects, we have no reason to draw any inference concerning
any object beyond those we have experience of.*[9]

Hume was of course aware that one cannot live one's life without constantly
employing this type of induction. But he argued that 'belief is more properly an
act of the sensitive than of the cogitative part of our natures'. The fact that we
base our lives on the belief that events are causally related does not imply that
there could be a rational demonstration that this is indeed so. Rational argument
has its limitations, like everything else.

It is difficult for a non-philosopher to develop a feeling for the force of
these arguments without considering some examples. Let us suppose that one
has a clock which ticks once every second for a year before running down,
but that one has forgotten when the battery was last changed. In this instance,
because we know something about its interior mechanism, we are quite unper-
turbed that the inductive inference is reversed. Each tick *reduces* the chances
of a further tick, because it takes us closer to the point at which the clock must
stop. This demonstrates that a naive belief in induction is not justified, and that
understanding the reason something happens provides much more compelling
grounds for believing in its continuation than any number of repetitions of the
event.

On closer examination the above argument merely opposes two different
types of regularity. The first is the regularity of the clock itself, while the second
is the regularity of the physical laws which govern the operation of the clock.
Our 'understanding' consists of preferring the regularity of the physical laws,
on the basis that this regularity has been tested very thoroughly in the past. If
we take this stand, then Hume reminds us that even if these laws have operated
exactly as we believe in the past, we have *no evidence at all* that the future will
resemble the past.

This is not mere philosophy. Consider the tale of the pig:

> *A pig is cared for by a farmer, and all of the evidence at its disposal indicates
> that the farmer is its friend. Indeed he has taken every care for its health and
> welfare since the day it was born. Yet one day the farmer comes into its pen
> and takes it out to kill it for profit.*

The point of this story is not merely that the particular pig was deceived about
the nature of the world, but that *every* pig is in the same situation! Unlike
sheep and cows, the almost universal fate of pigs is to be killed and eaten. We
assume that we are not in the same situation, and that our apparent progress
in understanding the world has some basis in fact, but we can never *know* that
this, or indeed anything, is so. In order to retain some use for the word 'know',
one needs to exclude the possibility of deliberate and systematic deception by
some super-being.

There have been many attempts to resolve the above problem. I rather like
the following one, even though it has an important flaw. As we have become
more sophisticated we have replaced a belief in the regularity of events, such

as the rising of the sun, by a belief in the continued validity of scientific laws. These laws have become ever more general, and we have discovered that all aspects of our bodily functions depend in a critical way upon the continued validity of a few very fundamental laws, which also control all other aspects of the world. Every proper use of induction can be reduced to the assumption that these laws will continue to operate. We are justified in believing this, because if the laws change, even slightly, we will never know it. We will already have died or even disintegrated.

The flaw is the assumption that a change in the laws of physics would have to apply everywhere simultaneously. It is *logically possible* that there might be occasional localized disruptions of physical law, which were beyond our understanding. This is indeed what miracles are supposed to be. The Moon might disappear for a day and then reappear and continue to orbit the Earth as if nothing special had happened. Such events, if common enough, would certainly make people have less confidence in the use of induction! Well, maybe, but it would hardly be sensible to plan our lives on the assumption that such events are about to become common.

Richard Swinburne has devised another defence of our use of induction. He argued that *as a rational being* one has no choice but to assume that the simplest explanation of a collection of facts is most likely the right one. If we have to make a real life decision about the next term in the sequence 2, 4, 6, 8, 10, 12, . . . then we will choose 14 because it is the value yielded by the simplest possible formula. If no other information is available people are right to prefer this choice to that provided by more complicated formulae. This thesis presupposes the possibility of agreeing on criteria of simplicity, which are not easy to formulate even though there is often a clear consensus about the relative simplicity of two rules. The above is not intended to be an argument about what is actually true. All Swinburne is claiming is that people are right to prefer the simplest theory, as in fact they do in all situations in real life.

One must be careful not to over-interpret the concept of likelihood. In the first half of the twentieth century Carnap tried to formalize the likely truth of a scientific theory within Kolmogorov's theory of probability. Both Carnap and Reichenbach were severely criticized by Popper in *The Logic of Scientific Discovery*, where he demonstrated why no such programme could succeed. Without following the details of the argument, one can see the one of the problems by considering the anomalous orbits of Uranus and Mercury. The apparent failure of Newton's laws to predict the motion of Uranus exactly led to a search for a new planet whose gravity might have been perturbing its orbit, and Neptune was discovered in 1846. On the other hand a similar search for a new planet to explain the slight failure of Newton's laws to predict the motion of Mercury led nowhere. Eventually Einstein's general theory of relativity was needed to resolve the matter. So here two fairly similar phenomena turned out to have completely different explanations. It strains credulity to argue that any philosophical study or probability calculus could have anticipated such developments.

In spite of Hume's scepticism, philosophers in the eighteenth and nineteenth centuries had to explain the fact that human beings appeared to have two forms of knowledge of the world which everyone agreed were not subject to question. These were Euclidean geometry and Newtonian mechanics. The first seemed to be a description of how physical space actually is, while the second seemed to give an exact description of how bodies move. Unfortunately even these certainties started to unravel in the second half of the nineteenth century. The process started when Riemann showed that there were an infinite number of different possible geometries, all of equal standing from a mathematical point of view. When Einstein's general theory of relativity was confirmed by observations of the bending of starlight during an eclipse in 1919 the implications could not be evaded. It was seen that both Euclidean geometry and Newton's laws of motion were just theories, perhaps extremely accurate ones, but no longer 'true' descriptions of the world. The advent of quantum theory a few years later further undermined people's belief that reality and any particular theory we may have of it are identifiable. Einstein was fully aware of the philosophical implications of his work, and frequently wrote along the following lines:

> In the previous paragraphs we have attempted to describe how the concepts space, time, and event can be put psychologically into relation with experiences. Considered logically, they are free creations of the human intelligence, tools of thought, which are created to serve the purpose of bringing experiences into relation with each other, so that in this way they can be better surveyed. The attempt to become conscious of these fundamental concepts should show to what extent we are actually bound to these concepts. In this way we become aware of our freedom, of which, in case of necessity, it is always a difficult matter to make sensible use . . .
>
> Why is it necessary to drag down from the Olympian fields of Plato the fundamental ideas of thought in natural science, and to attempt to reveal their earthly lineage? Answer: In order to free these ideas from the taboo attached to them, and thus to achieve greater freedom in the formulation of ideas or concepts. It is to the immortal credit of D. Hume and E. Mach that they, above all others, introduced this critical conception.[10]

In an attempt to eliminate the problem of justifying induction, Karl Popper proposed a different approach to the nature of scientific knowledge in 1934. Many years were to pass before the importance of his ideas was realized. This was partly because his book *Logik der Forschung* was not translated into English (as *The Logic of Scientific Discovery*) until 1959, and partly because his ideas were so at variance with the dominant tradition of logical positivism and linguistic analysis in Oxbridge at that time. He argued that scientific knowledge advances not by the application of the inductive process, but by the formulation of conjectures, which are then tested in the laboratory. A theory can never be proved by repeated testing, but can be refuted by failing some test. His idea that all scientific knowledge is provisional, and subject to possible later refutation and replacement by a new theory, provided an attractive explanation of developments in physics during the twentieth century, and became the new orthodoxy. Outside the circle of professional philosophers his ideas are often mentioned as if they alone are incapable of refutation!

Scientific theories can be tested in many different ways, and it is better to make a variety of quite different tests than a large number of similar ones. In this respect Popper's ideas correspond better to actual scientific practice than Hume's references to the 'constant conjunction of two objects'. The pig in my parable was deceived because it relied entirely upon the repetition of one event (its survival to the next day) and could not put this into a wider context. We are not deceived by the repeated ticking of a clock because we know about the law of conservation of energy, which has been tested in many quite unrelated situations.

Unfortunately, while Popper's ideas were taking root among scientists, a serious flaw in his theory was discovered by Hilary Putnam. He argued that it did not break free of the problem of induction as Popper intended. It might be possible to argue that Popper's theory applies to certain types of scientific work in a laboratory, but scientists do not only test conjectures, they eventually recommend that others rely upon the laws which they believe they have found. Putnam pointed out that if testing conjectures were all that scientists ever did, then science would be a wholly unimportant activity. The fact that a law has been highly corroborated, i.e. has never failed a critical test, is taken in the real world as evidence that it may be used in practical contexts. The lack of a logical basis for this is exactly the difficulty which Hume had pointed out.[11]

Although Popper claimed to be writing about empirical science in general, *The Logic of Scientific Discovery* never mentioned geology, biology, or evolution, but concentrated on physics, mathematics, logic, and probability. Indeed he defined a scientific theory as a universal statement consisting of symbolic formulae or symbolic schemata. So plate tectonics would not be a scientific theory by his definition. Popper's work has had considerable impact on physical scientists, but there are situations in which knowledge cannot sensibly be described as provisional, even though it is directly testable and wholly scientific. Indeed in a public lecture in Cambridge in November 1994 Max Perutz criticized Popper's theories as having no relevance to the way molecular biology and chemistry have developed. In support of this consider the following list of facts, which are no more provisional than our belief that the world is round. None of them was known four hundred years ago, none involves symbolic formulae and each required a considerable effort to discover. There are *many* other facts of a similar type.

- The Earth rotates about its axis and also orbits around the Sun.
- The blood circulates around the body, pumped by the heart.
- Diamond, graphite, and coal are all predominately composed of the same element, carbon.
- All material objects are composed of atoms.
- There is a close connection between electricity and magnetism.

- Malaria is caused by a parasite transmitted by mosquitoes, which takes up residence in people's red blood cells.
- Many insects, and in particular bees, have compound eyes.

Although the above statements are surely objectively true, that does not mean that they have to be accepted on authority. They remain testable, and hence *in principle* refutable; but experience tells us, as certainly as anything can, that they will pass any tests made of them. So it is with the statement that $3 \times 4 = 12$. You can test it for yourself by putting down four rows of three beans, and then counting them, but this does not mean that there is even a slight chance that the identity is wrong. Scientific knowledge may also be certain even when it is not expressed in words or equations. Robert Hooke's book *Micrographia*—published in English rather than the usual Latin—had an enormous and immediate impact when published in 1665: its many beautiful engravings of objects seen using a microscope changed the *kind* of questions people could ask about nature. Figure 10.2 is of the most famous flea in human history![12]

One might try to imagine the possibility that a future physics will have dispensed with the need to believe in atoms, and that our descendents will describe chemistry, molecular biology, genetics, solid state physics, and nuclear physics in some quite different manner. It is of course impossible to *prove* that this could not happen, but there is no historical case in which an idea with such diverse experimental support has been completely abandoned. The best

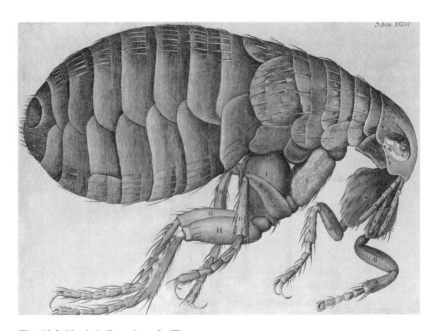

Fig. 10.2 Hooke's Drawing of a Flea
© The Royal Society

candidates are alchemy, phlogiston (discussed further on page 271), and the geocentric universe, but more on the grounds of their longevity than of the amount of evidence supporting them. An indication of the volume of current research supporting Dalton's atomic theory of chemical compounds is given by the following table, which lists the size of the journal *Chemical Abstracts* for three chosen years in the twentieth century. Size is measured by the combined thickness of the relevant volumes!

Year	Size
1920	21 cm
1955	44 cm
1990	318 cm

Similar statistics could be given for physics and mathematics, but the volume of current research in medicine and the life sciences is surely much larger.

One might try to rescue Popper by suggesting that he was writing about *theories* and that the above items are *hypotheses* or *facts*. However, this will not do, because there is actually no sharp distinction. Dalton's claim that material bodies are composed of atoms was certainly a theory which many chemists did not believe during the whole of the nineteenth century. At that time it satisfied the criterion of providing an abstract quantitative mathematical structure which explained a steadily increasing range of experimental facts, while being well beyond direct experimental verification. During the twentieth century the sheer *variety* of different types of corroboration of the theory has eventually made atoms a part of the very language which we use to describe phenomena. One can challenge individual items such as whether one can 'really see' single atoms in suitable microscopes, but collectively the weight of evidence is overwhelming.

On considering these examples a rather interesting fact emerges. Scientific theories can be true (or false in the case of phlogiston) when they can be expressed in common sense terms. On the other hand scientific theories which can *only* be expressed in terms of sophisticated mathematics have a much more provisional status. History shows that they may be replaced by better theories using entirely different mathematics at some future date. Mathematical theories do not *explain* the subject they refer to, but only provide models which enable predictions to be made. Recall Newton's own admission that there was something deeply mysterious in his theory of gravitation. Popper was right only in as much as he was referring to mathematical theories. They are simultaneously the most accurate theories we have and also the ones whose formulations have changed most with time.

It is an interesting fact that scientific theories are not discarded simply because they fail to predict a range of facts correctly. Paradoxical new discoveries prompt two responses. The first is an attempt to find errors in the new work, and the second is to try to incorporate the new facts into the existing theory with the fewest possible changes. Only if both of these responses fail do scientists start looking for a new theory which will explain all of the new facts in an economical and convincing manner. They never throw away the old theory until a superior replacement has been found, and frequently do not do so even then.

The current orthodoxy among physicists is that every (mathematical) theory has a domain of validity. Within its domain it yields reliable predictions, but as one approaches the boundary of the domain the predictions become steadily less accurate. Note that the idea of truth or falsity of theories is no longer an issue: all that remains is whether theories yield accurate predictions or not *within some domain*. This accounts for the peculiar fact that Newtonian mechanics continues to be used long after it has supposedly been superseded by general relativity and quantum theory. Its domain of applicability involves restricting the speeds of the bodies to be much less than the speed of light, and avoiding the scales of size and energy at which quantum effects become important.

The acceptance of the radically new quantum theory in the late 1920s resulted from a collective agreement that the new theory explained a wider variety of facts than did the previous one. The reason why the older Newtonian mechanics was not abandoned with the advent of quantum theory is quite simple. The equations of quantum theory might have been quite simple to write down, but they were extremely hard to solve in all except the simplest cases. Even though the old theory was less accurate, its much greater technical simplicity ensured that it would be used in all cases in which quantum effects were not of paramount importance. In particular Newton's theory, although 'superseded', is still the one used to calculate the orbits of satellites and the trajectories of spacecraft.

Empiricism versus Realism

In this section we consider some fairly recent contributions to the philosophy of science, which again illustrate how difficult it is to produce a satisfactory description of the scientific enterprise. The two major contenders are called scientific realism and empiricism.[13]

The issue in this debate is not whether the world exhibits regularities, which can be discovered and classified by scientists. This can hardly be denied. Faraday's work relating electricity and magnetism has transformed almost every aspect of our daily lives. The same can be said of Dalton's atomic theory. The Curies' laboratory study of radioactivity has had profound implications for the study of stellar evolution, geological dating, and several different industries.

During the nineteenth century many elements were heated and then examined using spectrometers, which split their light into its different colours. Sharp spectral lines were observed, with a different and characteristic pattern for each element. *Exactly* the same patterns are observed in the light received from very distant stars viewed using our most powerful telescopes. This provides convincing evidence that physics is the same over an enormous range of times and distances. There is no logical proof that laboratory observations should apply to the larger world, but nevertheless it has been so, time and time again. That, ultimately, is why science is more than a hobby for eccentrics.

The issue therefore is not the existence of regularities in the world, but the status of theories which describe those regularities. Let us start with realism, the natural position for any scientist. In his influential book *The Scientific Image* Van Fraassen defined it as follows:

> Science aims to give us, in its theories, a literally true story of what the world is like; and acceptance of a scientific theory involves the belief that it is true.[14]

He emphasized the inclusion of the words 'aims' and 'literally' in this definition. He also stated that if belief comes in degrees, so does acceptance, and we may then speak of a degree of acceptance involving a certain degree of belief that the theory is true. The above should be contrasted with his definition of empiricism:

> Science aims to give us theories which are empirically adequate; and acceptance of a theory involves a belief only that it is empirically adequate.

Van Fraassen is not a realist, and in particular puts the evidence provided by microscopes into the 'theoretical' category. Indeed he is only prepared to accept as real what can be perceived using the unaided senses. Needless to say there are many philosophers and scientists who disagree with him. Van Fraassen's position cannot be disproved on logical grounds, but logic is not everything. Figure 10.3 is of the skeleton of a marine protozoan called a radiolarian, magnified about a thousand times using a scanning electron microscope. Unlike fleas, the very existence of radiolarians is only known because of the existence of microscopes. Nevertheless, such images are so obviously similar in type to what we see with our naked eyes that to regard them as theoretical entities strikes me as extremely contrived. But it is not merely a matter of saying that 'seeing is believing'. Scientists regularly sort cells into types by moving them around with a probe, and inject material into single cells using extremely fine hollow needles. Such manipulations greatly increase one's confidence that there is something real at the other end of the microscope.

I am emphatically *not* saying that what we see using a microscope coincides exactly with what is there. Indeed much of Chapter 1 was devoted to showing that this 'naive realism' is not justified even when looking at objects of our own size. Different types of microscope provide different images, and one often has to work hard to find the best way of viewing almost transparent cellular bodies. But the same would be true for almost transparent objects of our own size.

Even if one accepts the reality of entities seen using a light microscope, the existence of viruses might be questioned. They are theoretical entities in the sense that we have to infer their existence from primary sense data. On the other hand the evidence is so overwhelming and comes from so many independent sources that it is impossible to imagine it being overthrown. I doubt that *any* scientist considers that the existence of viruses is open to serious debate. The situation is *not* comparable with the status of Newtonian mechanics in 1900. The amount of evidence for the existence of viruses is vastly greater and more varied than it ever was for Newtonian mechanics. Similar comments could be made about DNA, protein molecules, all of the fainter stars in our galaxy, other

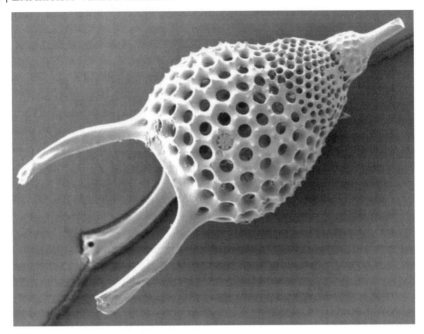

Fig. 10.3 A Radiolarian
Photographed by Tony Romeo, Electron Microscope Unit, The University of Sydney

galaxies, etc. The existence of all of these is inferred via the use of sophisticated instruments and not by direct observation. Nevertheless it would be absurd to regard their existence as merely hypotheses which lead to good predictions. A few philosophers might take this attitude to prove how open-minded they are, but they will convince nobody else.

The above comments should not be taken as an endorsement of all scientific theories. Each one has to be considered on its own merits. Some theories have now progressed to the status of settled fact, but there are many others which are much more provisional. If we look into the sky on a cloudless night we may see many randomly scattered points of light. As a result of a long chain of theoretical arguments involving optics, spectroscopy, and nuclear physics we now believe that these are caused by large glowing bodies at incredible distances. As this stage in the development of astronomy it would be absurd to deny the existence of stars, but many questions about them are still open. For example, the recent 'solution' of the solar neutrino problem may be correct, but it is still capable of being doubted. This being so, we cannot yet be *sure* that we fully understand the dynamics of our own sun, let alone all others.

Everything written above favours the realist interpretation of science. Unfortunately, once one turns to scientific theories of a highly mathematical kind the situation reverses. The three most successful such theories are Newtonian mechanics, quantum mechanics, and relativity theory. The *predictions* of Newtonian mechanics are extremely accurate for medium sized and slow moving bodies, but scientific realism is about truth, not predictive power. Appeals to predictive power are retreats to an empiricist position. Quantum mechanics and relativity theory depend upon utterly different views of 'what the world is like'. The Newtonian action at a distance has disappeared from general relativity, but is still present in quantum mechanics. The three theories are based on completely different fields of mathematics. So fundamental physics, as currently practised, fails van Fraassen's test for realism.

Let us look at a more recent definition of Psillos:[15]

> *Mature and successful scientific theories are well-confirmed and approximately true of the world. So the entities posited by them, or, at any rate, entities very similar to those posited, inhabit the world.*

Unfortunately Newtonian atoms are *not* very similar to quantum mechanical atoms: the latter can interfere with themselves in double slit experiments with measurable consequences; the former cannot. Scientists' understanding of the nature of protons was entirely changed by the discovery of quarks, so presumably the earlier very successful theory of protons was immature. Recent discoveries at the cutting edge of cosmology are just as amazing. Observations of the rotation of galaxies show that they must be surrounded by large quantities of dark (i.e. unobservable) matter, and even of unknown and exotic forms of energy. What we can observe may be only about 5% of what is there. It seems possible that the origin of the universe may involve a bizarre phenomena going under the name of inflation, and that there may be many parallel universes with which we can have no contact. Physicists cannot even agree on the nature of space-time: its preferred number of dimensions at present is ten (or possibly eleven). The inevitable conclusion is that fundamental physics and cosmology are not mature sciences. Psillos's definition can only be saved by agreeing that there are no mature sciences!

Since (these two) philosophers have not resolved the issue, perhaps they are approaching the problem from the wrong angle. Almost all physicists claim to be realists, so it might be worthwhile to see what one of them means by this term. We choose Steven Weinberg, who has written several very thoughtful popular books about science, and who has won a Nobel Prize for his work in fundamental physics. In *Dreams of a Final Theory* he writes:

> *My argument here is for the reality of the laws of nature, in opposition to the modern positivists, who accept the reality only of that which can be directly observed. When we say that a thing is real we are simply expressing a sort of respect. We mean that the thing must be taken seriously because it can affect us in ways that are not entirely in our control and because we cannot learn about it without making an effort that goes beyond our own imagination.[16]*

This is a rather weak concept of realism, but he continues with the following:

> *But I have to admit that my willingness to grant the title of 'real' is a little like Lloyd George's willingness to grant titles of nobility; it is a measure of how little difference I think the title makes.*

I have struggled to understand what he means by this sentence, without success. The rest of the passage suggests that he *may* be saying that scientists are not free to invent laws arbitrarily, and that it is not worth worrying about the absence of absolute criteria of truth, or of reality. This is certainly true from the point of view of physics, but philosophers of science are not trying to facilitate the progress of physics. They are trying to understand the *nature*, or *status*, of scientific knowledge, an entirely distinct matter. Developments in physics must be of importance to them, but that does not imply that their conclusions should have any relevance to the practice of physics. This is not a criticism, since exactly the same is true of many other people, for example computer chip manufacturers.

A part of the problem is that it is extremely hard to detach oneself from one's beliefs about the future endpoint of the scientific enterprise. New discoveries may lead science into quite unanticipated territory, as quantum theory did, destroying Laplacian determinism in the process. We may one day have a 'Theory of Everything', as Weinberg expects, and we may not. Arguing that it must necessarily exist, even if we never succeed in finding it, is simply expressing the prejudice that the world must inevitably be law-like, and that the laws must be mathematical. Seeking mathematical explanations is fine as a method of investigating the world. It has been extremely successful, but that does not commit one to declaring that there cannot be any other way of thinking about the world.

It seems best to be content with a description of science as it now is, and to attribute goals only to individual scientists. We might characterize science simply as the systematic enquiry into the properties and behaviour of the natural world. Scientists try to obtain detailed explanations of aspects of the world by using a combination of theoretical arguments and empirical tests. These must be accessible to everyone, subject to the acquisition of the necessary technical skills. The conclusions must be testable in the natural world and falsifiable. Pure mathematics is not science, nor is Christian Science.[17]

The Sociology of Science

A weakness of the last section is that it focuses on only one aspect of science. John Ziman, who describes himself as a lapsed physicist, has argued for many years that we can only understand science fully by considering also its psychological and social aspects.[18] The psychological aspect refers to the process of discovery by individual scientists, who now increasingly work in organized teams. Any detailed investigation of this must be based on comparing their note books and personal accounts. Every historian knows that the latter are frequently unreliable: people unconsciously simplify the process of discovery and often do not mention false trails; after a period of years they forget the sequence of events, and occasionally deliberately misrepresent them.

The social aspect of science refers to the process by which the community adopts a new discovery. This is frequently highly complicated and drawn out, as well as being dependent upon the reputations and personalities of the scientists involved. Textbooks regularly ignore this aspect of science, or reduce it to a parody which does not mention any of the doubts expressed at the time. The importance of these social issues came to the attention of the general public when Thomas Kuhn published his book *The Structure of Scientific Revolutions* in 1962.[19] He argued that science does not consist only of the steady accumulation of knowledge, each bit carefully considered by the community and then added onto the list of established facts. This he called normal science, to be contrasted with revolutionary science. In the latter there is a major change of viewpoint, which he called a paradigm shift, in which the old framework is demolished and replaced by a radically new one. Two obvious examples of theories which caused major paradigm shifts are Darwin's theory of evolution and Einstein's general relativity. Kuhn argued that scientists do not make decisions about what paradigm to accept on narrowly logical grounds. Like everyone else, they depend upon judgement and experience.

Kuhn's emphasis on discontinuities in scientific development and on changing paradigms were valuable contributions to the understanding of science, as was his declaration that science does not proceed only by logical arguments. Indeed one could go further: *most* innovative research depends as heavily upon judgements as upon logic, and leads to changes of attitude towards what was previously known. Sometimes these are large changes justifying the term paradigm shift, and sometimes they are small. There is no principled way of distinguishing between normal and revolutionary science.

Kuhn also introduced the notion of incommensurability between theories. This is difficult to explain, and we will start with one of the many examples which Kuhn described. This involves phlogiston, a word referring to a substance (or principle) whose existence was widely accepted in eighteenth century chemistry.[20] It has no direct translation into modern chemistry, and we now consider that no such substance exists. But with sufficient effort one can understand how eighteenth century chemists built up a coherent picture of chemical phenomena using phlogiston. Most explanations of phenomena using phlogiston can be interpreted in modern chemical language, but the interpretation is very different from case to case. Kuhn described phlogistic chemistry and modern chemistry as incommensurable.

The concept of incommensurability was widely discussed, and even criticized as ultimately incoherent. Later in his life Kuhn claimed that many of his critics were over-interpreting what he had written; he also accepted some responsibility for the misunderstandings.[21] Let us start with the strong incommensurability position he is widely considered to have been advocating in *The Structure of Scientific Revolutions*. This stated that there could be no basis for rational discussions between the advocates of a sufficiently radical new theory and its predecessor. The concepts they used were so different that one simply had to adopt one of the two world views. Scientific revolutions were rather like political ones, the winner being determined by a power struggle rather than by rational argument.

This strong version of incommensurability was criticized by several of Kuhn's contemporaries, and is not sustainable. As a test case let us consider the birth of quantum theory in 1925–26. This must be counted as one of the greatest revolutions on scientific thought ever: it has totally transformed physics, chemistry, and even biology over the last century. The scientific community agreed fairly quickly that it was a huge advance on Newtonian mechanics, even if some, such as Einstein, expected that it would eventually be replaced by something less mysterious. However, it is not strongly incommensurable with Newtonian mechanics: the earlier theory is derivable as a limit of the new one as Planck's constant converges to zero; in the reverse direction many of the notions of Newtonian mechanics (space, time, potential energy, kinetic energy, momentum) have direct analogues in quantum theory. While quantum theory undoubtedly reigns in the micro-world of atoms, this is not to say that quantum theory and Newtonian mechanics make the same predictions about the behaviour of all sufficiently large bodies: superconductivity and superfluidity are among the macroscopic effects which are inexplicable along Newtonian lines. The point is rather that quantum theory enables scientists to work out when Newtonian mechanics will make the right predictions.

The same applies in to the discovery of the structure of DNA, and to plate tectonics, the theory explaining continental drift. Each was entirely comprehensible within the existing frameworks of the two subjects. They provided detailed mechanisms to explain a range of phenomena which had previously not been understood. Both were confirmed by rapidly increasing amounts of evidence, which caused huge changes to sweep through the subjects. They might have shown that certain earlier beliefs were *wrong*, there is no evidence of (strong) incommensurability of paradigms.

In 1976 Kuhn wrote that he intended incommensurability between theories to be interpreted in a much weaker sense: he agreed that two incommensurable theories could be compared, even though some terms used in one might have no analogues in the other.[22] Exploring the extent to which he changed his views with time, as opposed to simply finding a clearer way of expressing them, is far beyond the scope of this book. But Kuhn's ideas about incommensurability have been so influential (often unfortunately so), that I must quote the following passage of his:

> *To name persuasion as the scientist's recourse is not to suggest that there are not many good reasons for choosing one theory rather than another. It is emphatically not my view that 'adoption of a new scientific theory is an intuitive or mystical affair, a matter for psychological description rather than logical or methodological codification' ... What I am denying then is neither the existence of good reasons nor that these reasons are of the sort usually described. I am, however, insisting that such reasons constitute values to be used in making choices rather than rules of choice. Scientists who share them may nevertheless make different choices in the same concrete situation ... In such cases of value conflict (e.g. one theory is simpler but the other is more*

accurate) the relative weight placed on different values by different individuals can play a decisive role in individual choice.[23]

To the extent that he is supporting reason and judgement, as opposed to fixed rules, for the acceptance of theories, this seems entirely acceptable.

Kuhn's ideas have had a major impact on the development of the sociology of science. This is distinguished from the history and philosophy of science by its methodology. At its best it merely avoids addressing the truth of scientific beliefs, considering *only* the social issues. This is entirely defensible: sociologists studying an immature and controversial field are not in a position to judge which view will eventually prevail. Their accounts of personal and group interactions would hardly be trustworthy if they had prejudged the outcome of future experiments.

According to Barnes, Bloor, and Henry the pursuit of science has more in common with other forms of human activity than is commonly admitted:

> *In no way does this imply any criticism of science, but it does suggest that the realisms of science are particular instances of something common to all forms of culture and implicit in all forms of practice. The presumed reality of ghosts and spirits organizes life in many tribal cultures ... Scientists are distinctive in the theoretical objects they currently assume to be 'really there'. But in their sense that such objects are there, and in their use of the techniques and devices of the realist mode of speech, they are typical of human beings in all cultures.*[24]

Unfortunately there are those who go far beyond the above statement, claiming that Western Science is merely one culture among many, and that one has the right to reject its conclusions if one feels uncomfortable with them. Richard Dawkins claims to have been confronted with this type of attitude regularly when speaking about the theory of evolution in public meetings. His response is now famous: 'show me a cultural relativist at thirty thousand feet and I'll show you a hypocrite'. It is easy to refute the extreme position adopted by such people. That objects thrown upwards return to Earth is a fact about which one can hardly argue, and given that, the relationship between the force with which one throws an object and the length of time before it returns is not a matter on which one can have a variety of opinions. All available evidence confirms that the same scientific laws operate everywhere and that no changes in the fundamental physical constants have taken place over billions of years.

For a more systematic criticism of cultural relativism I recommend the recent book *The Truth of Science* by Roger Newton. Accepting his arguments does not, however, render irrelevant other matters raised by more serious sociologists of science. They have examined how scientific discoveries were actually made in a number of case studies, and have found that the process was nothing like the 'official' method by which science is carried out. The case of the measurement of the charge of the electron by Millikan provides a very clear example of this. Barnes, Bloor, and Henry pointed out that in his 1913 paper in Physical Review, Millikan used only 58 of the 175 experiments which he performed according to his notebooks; the others were rejected because he regarded them as anomalous in some respect. This might have resulted in his being accused of falsification if his results had not stood the test of time. They also describe the contemporaneous experiments of Ehrenhaft, which yielded quite different results for the electron charge, and were eventually eliminated from the corpus of science without being convincingly refuted.

Our current beliefs about the charge of electrons are not based upon the results of either of these scientists. Millikan realized that direct measurements of the charge were in fact possible. Since that time technology has progressed so far that his experiments would now be regarded as extremely primitive. We remember him rather than Ehrenhaft because the value he obtained turned out to be more or less right. Much more accurate measurements are now possible and have been verified by their consistency with a wide variety of other knowledge.

At the start of the seventeenth century Francis Bacon declared that scientists proceeded, or should proceed, by collecting facts until they saw the pattern into which they fell. There are only a few examples which fit this description. Kepler did indeed study Brahe's astronomical data for years, before he concluded on this basis alone that the orbits of the planets were elliptical. Much more recently a lot of progress in genetics depends upon comparing the genomes of very different species, using the massive databases which have been collected.

Particle physics is at the opposite extreme. Experiments are designed on the basis of a large amount of theoretical calculation, in order to test a specific prediction in a true Popperian manner. In this field the observational data are so far removed from the objects they are supposed to be related to, that the very existence of the objects might well be questioned: detecting an atom of argon in an underground vat rich in chlorine is hardly the same as seeing a solar neutrino. The interpretation of such experiments depends on several layers of theory, but theory which is better accepted than that being investigated. It may take decades before scientists reach a consensus about the validity of a new theory. There are no rules setting out what criteria should be used to decide such questions: every case is unique and can only be settled by a combination of experiment and informed debate.

Most of the general public were not aware of this process until recently because scientists have had an interest in portraying their subject as objective and free of the confusions which are all too obvious in most aspects of life. Their increasing willingness to join in public debates and to admit their ignorance is very much to be welcomed. If only governments were willing to spend more on research *before* the regular disasters which hit us. Alas, the time scale of politicians is reckoned in weeks or months, not the years, or even decades, which scientists need to make reliable judgements.

Science and Technology

One of the grounds for claiming that science is not morally neutral is the impossibility of separating the scientific enterprise from applications which some regard as unethical or ecologically unsound. It cannot be denied that much of the scientific enterprise is an attempt to control the world rather than to understand it. The development of accurate timepieces was a response to the need for reliable navigation around the world. The science of thermodynamics arose in the nineteenth century out of the need to make more efficient steam engines. Discoveries of peculiar effects in quantum theory have had such important consequences for the growth of the semiconductor/computer industry that it is difficult to disentangle the two. Much of the development of chemistry has been a response to industrial needs, from the synthesis of fertilizers onwards.

Scientists frequently defend controversial areas of research on the grounds that we must have the abstract knowledge to enable us to make informed choices about how society should develop. This presupposes that a distinction between science and technology can be made. Traditionally this was quite easy: science was what was done in universities while technology was what was developed by industries. Indeed there are still cases in which one can be clear that a subject is science and not technology. Particle physics, astronomy, cosmology, and the study of human origins spring to mind. Once upon a time this distinction was also clear to the British research councils: they funded fundamental science and expected industry to fund its applications. However, times have changed, and the new orthodoxy is to expect university researchers to pursue lines of research which will benefit the economy, and to set up spin-off companies to do so.

There is plenty of contemporary evidence to support the claim that scientific judgements can be contaminated by political considerations. The spread of the disease BSE in the UK in the 1980s was a result of the adoption of unsafe livestock feeding practices. According to a British Government report in 2000, the behaviour of the Government of the time had been designed more to allay fear than to provide full information about the possible risks.[25] Two of its many recommendations—'epidemiologists, particularly those in the public sector, should make available the data upon which their conclusions are based' and 'an advisory committee should not water down its formulated assessment of risk out of anxiety not to cause public alarm'—speak volumes about the general Government custom of manipulating information. At least one Ministry (MAFF) was more concerned to protect the profits of the farming community than the health of the public it supposedly served.

A direct result of this scandal in the UK has been the collapse of public confidence about advice concerning related issues, such as the safety of GM foods. The public are behaving perfectly rationally in this matter. Although very few can assess the scientific issues personally, everybody knows that reassurances by those in authority are unreliable (some would say worthless) when the profits of major corporations or industries are at stake. The only way forward is for everyone to admit that no human activity can be 'safe' in an absolute sense. According to Robert May[26] 'the full messy process whereby scientific understanding is arrived at with all its problems has to be spilled out into the open'. The public must be allowed to compare the risks of any action with the benefits and with the risks of alternative courses of action.

The most recent controversy among the increasing number which science is presenting us concerns the cloning of human organs. For some this raises the horrors of Frankenstein's monster and is clearly unethical. For others it provides the possibility of saving the lives of those suffering from severe and irreversible illnesses. Those in favour argue that we should first determine the scientific possibility of producing cloned organs so that we can then make the ethical decisions about their use in a properly informed manner. Those opposed say that the large investments involved in such research always lead to its eventual use, and the decision not to proceed should be taken now. Public discussion is urgently needed, and is happening in the UK, but the number of totally novel and important issues appearing is overwhelming our capacity to discuss them seriously.

Conclusions

The distinction between science and technology is just one of many which we impose upon the world in order to organize our thoughts. Many other distinctions have been thought valuable by some but criticized by others. Thus Descartes constructed a dualistic philosophical system in which mind and matter were largely separated. Others such as Popper and Penrose have argued that the world has *three* aspects, the third being the world of human constructs for Popper and the Platonic world for Penrose.

One could equally well propose that the world should be divided into *five* categories: matter, fields, individual consciousness, information, and culture. Some people probably regard electromagnetic, gravitational, and quantum fields as being just aspects of matter, but this would not have been understood by Descartes. Penrose appears to think that consciousness might one day be amalgamated into the fields category in some speculative approach to quantum gravity. Functionalist philosophers try to absorb consciousness into the information category. I have argued that Platonism arises by trying to objectify human concepts, as Popper is close to doing with his third world. The right response to all of these theories is to remember that *we* choose the categories in order to organize *our* thoughts. They may be useful, even for centuries, but it is not plausible that any small number of categories can provide a comprehensive way of analysing the world.

At the present time the various sciences may be divided into two categories: those which depend heavily upon mathematics and those which do not. The former are generally regarded as more fundamental, but that does not imply that they have more significance for our everyday lives. Indeed, I do not know anyone who considers that the most fundamental branches of physics will have *any* technological relevance to our lives; the huge energies at which the effects become important more or less rule out basing industries upon them. On the other hand, there are many examples of important scientific developments which did not depend significantly upon mathematics. In particular, Galileo's astronomical discoveries shook the scientific world in spite of the fact that they depended upon nothing more than simple observation through a telescope. Linnaeus' systematic classification of organisms depended only upon observation aided by the use of the microscope. Darwin's *Origin of Species* does not contain a single equation. Faraday, the discoverer of electromagnetic theory and one of the greatest experimenters of all time, had an aversion to mathematics and rarely used more than simple ratios to describe his results. The recent theory of plate tectonics is completely comprehensible without any recourse to mathematics.

If one examines technology rather than science, one finds that mathematics (as opposed to simple arithmetic) had little effect on its development before the nineteenth century. An example is the design of ever more sophisticated locks over the last four centuries. This depended upon a high degree of three-dimensional geometric imagination and ingenuity, of the type which mathematicians pride themselves on possessing. However, no mathematics was involved, nor was any input from Newtonian mechanics, in spite of the fact that locks are mechanical devices. No doubt one could produce a mathematical specification of a combination lock if one tried hard enough, but it would be about as useful as describing a beautiful sunset in terms of the frequencies of the radiation involved.

When people talk about the unreasonable effectiveness of mathematics in the description of nature, they are usually referring to physics. The essence of this subject is to identify elementary systems whose behaviour depends upon an extremely small number of relevant factors, each of which can be expressed in numerical terms. This way of looking at the world was the responsibility of Galileo, more than any other single person. Once enough such systems have been understood, it is natural to start combining them in ever more complex ways, until, eventually, the mathematics is no longer able to cope, because of chaos, self-organized criticality, or some such issue. The power of mathematics only seems astonishing because of our lack of historical perspective. The subject today is the cumulative result of intense efforts by some of the most able people in the world over a period of two and a half thousand years. The acceleration of its progress in the seventeenth century was partly the result of Gutenberg's introduction of the printing press in the middle of the fifteenth century. Its relevance to the description of the world is a consequence of the fact that much of it was created for precisely this purpose.

The existence of major metropolises provides a different type of evidence of the astonishing achievements possible when whole societies cooperate, even unconsciously, over several centuries. New York was founded almost four hundred years ago, as New Amsterdam, and now boasts millions of inhabitants and a network of buildings, roads, and trains which would have been inconceivable to its founders. How much more one might expect to achieve in two and a half thousand years!

Let us return to mathematics. The physicist Roger Newton supports a less conventional view of its status than is usual among physicists:

> *It is not that Nature's own language is mathematics—as Galileo thought— and that we are thus compelled to learn every obscure rule and usage of that tongue to comprehend it, but that mathematics is* our *most efficient and incisive instrument for rational understanding of relations between things. If mathematicians have already built, with great ingenuity, elaborate structures containing results of long and hard thought, if they have devised concepts appropriate for reaching their conclusions, then scientists are only too happy to make use of this 'wonderful saving of mental operations'.*[27]

This makes much more sense than the seventeenth century view that God decided to create a universe which was governed by mathematical equations. This view of the world arose within the Judaeo–Christian tradition, but probably owed more to the (pre-Christian) Greeks than it did to Christianity. For many scientists it has now been replaced by the view that we construct models and theories about reality in order to help us to understand it.

The unfathomable mysteries of quantum theory provide forceful support for this more modest view of our abilities. We have used mathematics as a tool in helping us to understand the universe. A crucial assumption is that we can break the problem of understanding the universe up into small components which can be understood separately, and then combined to produce the big picture. This latter method has had brilliant successes, but it is fundamentally wrong. I have produced examples from Newtonian mechanics and quantum theory which demonstrate that the universe is an integrated whole, in which remote events can have rapid and important consequences for the behaviour of small systems, such as our brains. We, or if you prefer our, brains can then change the behaviour of other inanimate systems. The problem is not that the mathematics is wrong, but that it cannot be applied to sufficiently complex open systems, and such systems include *almost everything* which we encounter in our daily lives. To maintain that the relevant equations are correct, but that they are far too complicated and unstable to be soluble by us, is to adopt a philosophical stance. If one refuses to take this easy and conventional step, then the existence of free will becomes just one among many things which cannot be explained using equations. We only consider it to be uniquely difficult because we have made grossly exaggerated claims about our ability to understand the rest of reality. In fact we are merely intelligent apes, with an almost infinite capacity for self-congratulation. We are certainly far smarter than anything else around, but the fact that an elephant has a much longer trunk than anything else does not imply that it can suck up the ocean, whatever it may think about the matter.

There is no a priori reason why theories must necessarily be ordered in a hierarchy, the most fundamental being the ultimate and true explanation of everything. The evidence that our consciousness has an influence on our bodies as well as the other way around indicates that this is not a full explanation of the world. A more modest claim is that every scientific theory has ultimately to be consistent with reality, at least to some extent, if it is to be useful. The real world does not fall into neat divisions between physics, chemistry, biology, etc. These are boundaries which we impose upon it in order to break up its complexities into chunks which our brains can cope with. When we look at the structures of molecules, for example, we see that they lie on the boundary between quantum theory and chemistry. Both subjects claim to describe molecules, so the theories have to agree with each other if they are to agree with reality. Any improvements in the fit of either theory to the real world forces an increased similarity between the two theories. The degree of overlap will increase as the predictive powers of the two theories both improve. The claim that chemistry can be derived from physics is superficially attractive, but the historical chain of implication went in the reverse direction. The main evidence for atomic theory throughout the nineteenth century came from chemistry, just as the main evidence for the existence of genes came from the biology of whole organisms. Even today it is not possible to analyse the structure of large molecules without a major input from chemistry. We have seen that an *ab initio* approach to this problem will probably never be possible.

The situation in mathematics is similar. Gödel showed that this cannot be built in an orderly hierarchical manner starting from firm foundations. Hilbert's programme of deriving the whole of mathematics from a perfectly secure formal system is dead, but the bulk of mathematics *necessarily* survived this catastrophe because it had never been based upon this type of formalist reduction. Mathematics is a collection of overlapping fields, each with its own methodology. The reason that these are consistent with each other is that each is built using logic and geometrical imagination. Neither of these is infallible in our hands, because we are finite creatures, but experience shows that if apparent inconsistencies are examined slowly and carefully they can eventually be resolved.

Science may be regarded as providing a series of different views of the world, some sharp but narrow in scope and others broader but fuzzier. Each of the windows gives an equally valid view of different aspects of the same reality. As we study each view, we gradually sharpen our focus and find similarities with the views through other windows. The full complexity of reality is far beyond our ability to grasp, but our limited understanding has given us powers which we had no right to expect. There is no reason to believe that we are near the end of this road, and we may well hardly be past the beginning. The journey is what makes the enterprise fascinating. The fact that the full richness of the universe is beyond our limited comprehension makes it no less so.

Notes and References

[1] Bentley and Humphreys 1962

[2] No relation to the author of this book!

[3] An exhaustive study of the literature up to 1986 is given in Barrow and Tipler 1986.

[4] Polkinghorne 1998, p. 92

[5] Leslie 1989

[6] Barrow and Tipler 1986, Rees 1995, Rees, 1997, Rees 1999

[7] Klee, 2002

[8] Livio et al. 1989

[9] It is best to interpret the word 'object' as including events.

[10] Einstein 1982c

[11] See Putnam 1991. In the same article Popper rejects Putnam's criticism as completely misconceived, and Putnam rejects Popper's rejection. Philosophy is never dull!

[12] The drawing might well have been by Christopher Wren rather than by Hooke himself.

[13] My own position takes something from each of these, and is closest to entity realism. [Clarke 2001]

[14] Fraassen 1980

[15] Psillos 2000

[16] Weinberg 1993, p. 35

[17] Judge Overton made a definitive judgement to this effect in 1982 in the case of McLean v. Arkansas Board of Education. I have adopted parts of his statement of the essential characteristics of science above.

[18] Ziman 1995

[19] The French philosopher Gaston Bachelard anticipated some of the ideas of Kuhn, but his work is largely unknown in the English-speaking world.

[20] Kuhn 2000, p. 40–44

[21] Kuhn 2000, p. 155

[22] Kuhn 2000, p. 189

[23] Kuhn 2000, p. 157

[24] Barnes et al. 1996, p. 84

[25] Phillips 2000

[26] The current President of the Royal Society

[27] Newton 1997, p. 140

Bibliography

F. Acerbi: Plato: Parmenides 149a7-c3. A proof by complete induction? *Arch. Hist. Exact Sci.* **55** (2000) 57–76.

M. Agrawal et al.: PRIMES is in P. Preprint 2002.

P. W. Anderson: Shadows of Doubt. book review. *Nature* **372** (1994) 288–9.

P. W. Anderson: Essay review. Science: A 'Dappled World' or a 'Seamless Web'? *Stud. Hist. Phil. Mod. Phys.* **32**, no. 3 (2001) 487–94.

R. Ariew: Discussion. The Initial Response to Galileo's Lunar Observations. *Stud. Hist. Phil. Sci.* **32** (2001) 571–81.

M. Balaguer: *Platonism* and *anti-Platonism in Mathematics*. Oxford Univ. Press, Oxford, 1998.

G. W. Barlow: *The Cichlid Fishes: Nature's Grand Experiment in Evolution.* Perseus Publ., Cambridge, MA, 2000.

B. Barnes, D. Bloor, J. Henry: *Scientific Knowledge, a Sociological Analysis.* Athlone Press, London, 1996.

J. D. Barrow and F. J. Tipler: *The Anthropic Cosmological Principle.* Clarendon Press, Oxford, 1986.

P. Benacerraf: What numbers could not be. pp. 272–94 in *Philosophy of Mathematics*, eds. P. Benacerraf, H. Putnam, Second Edition. Cambridge Univ. Press, Cambridge, 1983.

P. Benacerraf, H. Putnam, eds.: *Philosophy of Mathematics*, First Edition. Basil Blackwell, Oxford, 1964.

W. A. Bentley and W. J. Humphreys: *Snow Crystals.* Dover Publ. Inc., New York, 1962.

P. Bernays: On Platonism in Mathematics. pp. 274–86 in *Philosophy of Mathematics*, eds. P. Benacerraf, H. Putnam, First Edition. Basil Blackwell, Oxford, 1964.

P. Bernays: The philosophy of mathematics and Hilbert's proof theory. pp. 234–65 in *From Brouwer to Hilbert. The Debate on the Foundations of Mathematics in the 1920s.* ed. P. Mancosu, Oxford Univ. Press, Oxford, 1998.

A. Bird: *Thomas Kuhn.* Acumen Publ. Ltd., Chesham, Bucks, 2000.

G. J. Black, P. D. Nicholson, P. C. Thomas: Hyperion: rotational dynamics. *Icarus* **117** (1995) 149–61.

L. Blum, M. Shub, S. Smale: On a theory of computation and complexity over the real numbers. *Bull. Amer. Math. Soc.* **21** (1989) 1–46.

R. S. Bradley: *Paleoclimatology.* Second edition, Academic Press, San Diego, London, Boston, 1999.

B. Brezger et al.: Matter-wave interferometer for large molecules. *Phys. Rev. Lett.* **88** (2002) 100404.

S. G. Brush: *The Kind of Motion We Call Heat, Book 1*. North-Holland, Amsterdam, 1976a.

S. G. Brush: *The Kind of Motion We Call Heat, Book 2*. North-Holland, Amsterdam, 1976b.

B. Butterworth: *The Mathematical Brain*. Macmillan, London, 1999.

N. Cartwright: *The Dappled World, A Study of the Boundaries of Science*. Camb. Univ. Press, Cambridge, 1999.

D. J. Chalmers: *The Conscious Mind*. Oxford Univ. Press, Oxford, 1996.

C. S. Chihara: *Constructibility and Mathematical Existence*. Clarendon Press, Oxford, 1990.

A. Church: Review of Turing. *J. Symbolic Logic* **2** (1936–7) 42.

P. M. Churchland and P. S. Churchland: *On the Contrary, Critical Essays 1987–1997*. The MIT Press, Cambridge, Mass., 1998.

S. Clarke: Defensible territory for entity realism. *Brit. J. Phil. Sci.* **52** (2001) 701–22.

P. J. Cohen: Comments on the foundations of set theory. pp. 9–15 in Axiomatic Set Theory, *Proc. Symp. Pure Math.* **Vol. XIII**, Part I. Amer. Math. Soc., Providence, RI, 1971.

I. B. Cohen, A. Whitman: A Guide to Newton's Principia. In *Isaac Newton: The Principia, A New Translation*, Univ. of California Press, Berkeley, 1999.

S. Colton: Refactorable numbers—a machine invention. *J. Integer Seq.* **2** (1999), Article 99.1.2.

M. C. Corballis: *The Lop-Sided Ape, Evolution of the Generative Mind*. Oxford Univ. Press, Oxford, 1991.

J. Cottingham: ed., *The Cambridge Companion to Descartes*. Camb. Univ. Press, Cambridge, 1992.

F. Crick: *The Astonishing Hypothesis*. Simon and Schuster, London, 1994.

J. Davidoff et al.: Colour categories in a stone-age tribe. *Nature* **398** (1999) 203–04.

E. B. Davies: Constructing infinite machines. *British J. Phil. Sci.* **52** (2001) 671–82.

E. B. Davies: Empiricism in arithmetic and analysis. *Phil. Math.* (3) **11** (2003) 53–66.

R. Dawkins: *The Extended Phenotype*. Oxford Univ. Press, Oxford, 1983.

H. W. de Regt and D. Dieks: *A contextual approach to scientific understanding*. Preprint, 2002.

S. Dehaene et al.: Abstract representations of numbers in the animal and human brain. *Trends in Neurosciences* **21** (1998) 355–61.

M. Dummett: Wittgenstein's philosophy of mathematics. pp. 491–509 of *Philosophy of Mathematics*, eds. P. Benacerraf, H. Putnam, First Edition. Basil Blackwell, Oxford, 1964.

M. Dummett: *Truth and Other Enigmas*. Duckworth, London, 1978.

C. Dunmore: Meta-level revolutions in mathematics. pp. 209–25 in *Revolutions in Mathematics*, ed. D. Gillies, Clarendon Press, Oxford, 1992.

S. Dürr et al.: Origin of quantum mechanical complementarity probed by a 'which-way' experiment in an atom interferometer. *Nature* no. 6697, vol. **395** (1998) 33–7.

J. Earman, J. D. Norton: Infinite Pains: The Trouble with Supertasks. pp. 231–61 in *Benacerraf and his Critics*. eds. A. Morton and S. P. Stich. Blackwell Publ., Cambridge MA, 1996.

G. M. Edelman: Memory and the individual soul: against silly determinism. Ch. 13 in *Nature's Imagination, the Frontiers of Scientific Vision*, ed. J. Cornwall, Oxford Univ. Press, Oxford, 1995.

G. D. Edgecombe (ed.): *Arthropod Fossils and Phylogeny*. Columbia Univ. Press, New York, 1998.

H. Eichenbaum: The topography of memory. *Nature* **402** (1999) 597–99.

A. Einstein: lecture delivered to the Prussian Academy of Sciences, January 1921. Taken from *Ideas and Opinions*, p. 233, Crown Publ. Inc., New York, 1982a.

A. Einstein: lecture delivered in Oxford, June 1933. Taken from *Ideas and Opinions*, p. 271, Crown Publ. Inc., New York, 1982b.

A. Einstein: from 'Relativity, the Special and the General Theory: a Popular Exposition', 1954. Taken from *Ideas and Opinions*, p. 364, 365, Crown Publ. Inc., New York, 1982c.

W. Ewald (ed.): *From Kant to Hilbert: a Source Book in the Foundations of Mathematics*. Clarendon Press, Oxford, 1996.

R. Feynman: *The Feynman Lectures on Physics, Volume 3*. Addison-Wesley, Reading, Mass., 1965.

R. Feynman: *The Character of Physical Law*. Penguin Books, London, 1992.

J. Fodor: Similarity and symbols in human thinking. *Nature* **396** (1998) 325–6.

B. C. van Fraassen: *The Scientific Image*. Clarendon Press, Oxford, 1980.

V. Frette et al.: Avalanche dynamics in a pile of rice. *Nature* **379** (1996) 49–52.

D. Gale: The truth and nothing but the truth. *Math. Intelligencer* **11** (1989) 65.

Galileo Galilei: *Dialogue concerning the Two Chief World Systems*, transl. by S. Drake 1967. Univ. of Calif. Press, Berkeley.

A. Garfinkel: Reductionism. Ch. 24 in *The Philosophy of Science*, eds. R. Boyd, P. Gasper, J. D. Trout. MIT Press, Cambridge, MA, 1991.

D. C. Geary: Reflections of evolution and culture in children's cognition. *American Psychologist* **50** (1995) 24–37.

M. Gell-Mann: *The Quark and the Jaguar*. Little, Brown and Co., London, 1994.

D. A. Gillies: Intuitionism versus Platonism: a 20th century controversy concerning the nature of numbers, in *Scientific and Philosophical Controversies*, ed. Fernando Gil, Lisboa, Fragmentos, 1990.

D. A. Gillies: An empiricist philosophy of mathematics and its implications for the history of mathematics. pp. 41–57 in Emily Grosholz and Herbert Breger (eds.) *The Growth of Mathematical Knowledge*, Synthese Library, Volume **289**, Kluwer, 2000a.

D. A. Gillies: *Philosophical Theories of Probability*. Routledge, London, 2000b.

Y. Goldvarg and P. N. Johnson-Laird: Illusions in modal reasoning. *Memory and Cognition* **28** (2000) 282–94.

R. F. Hendry: Models and approximations on quantum chemistry. *Poznan Studies in the Philosophy of the Sciences and the Humanities* **63** (1998) 123–42.

R. F. Hendry: Molecular models and the question of physicalism. *Intern. J. Philos. Chem.* **5** (1999) 117–34.

W. Hirstein, V. S. Ramachandran: Capgras syndrome: A novel probe for understanding neural representation of the identity and familiarity of persons. *Proc. Roy. Soc. London B* **264** (1997) 437–44.

W. Hodges: Turing's Philosophical Error? Ch. 6 in *Concepts for Neural Networks*, eds. L. J. Landau, J. G. Taylor, Springer-Verlag, London, 1996.

D. Hoffman: *Visual Intelligence: How We Create What We See*. W W Norton, New York, London, 1998.

D. R. Hofstadter: *Gödel, Escher, Bach: An Eternal Golden Braid*. Penguin Books, London, 1980.

M. Hogarth: *Predictability, Computability, and Spacetime*. Ph D thesis, University of Cambridge, June 1996.

S. Hollingdale: *Makers of Mathematics*. Penguin Books, London, 1989.

T. H. Huxley: *Evolution and Ethics and Other Essays*. Macmillan and Co. Ltd., London, 1894.

E. T. Jaynes: Bayesian methods: general background. pp. 1–25 in *Maximum Entropy and Bayesian Methods in Applied Statistics*, ed. J. H. Justice, Camb. Univ. Press, 1985.

T. C. Johnson et al.: Late Pleistocene dessication of Lake Victoria and rapid evolution of cichlid fishes. *Science* **273** (1996) 1091–3.

R. Klee: The revenge of Pythagoras: how a mathematical sharp practice undermines the contemporary design argument in astrophysical cosmology. *Brit. J. Phil. Sci.* **53** (2002) 331–54.

T. S. Kuhn: *The Road since Structure. Philosophical Essays, 1970–1993, with an Autobiographical Interview*. Univ. Of Chicago Press, Chicago, 2000.

I. Lakatos: *Mathematics, Science and Epistemology. Philosophical Papers vol. 2*. Cambridge Univ. Press, Cambridge, 1978.

L. J. Landau: Penrose's Philosophical Error. Ch. 7 in *Concepts for Neural Networks*, eds. L. J. Landau, J. G. Taylor, Springer-Verlag, London, 1996.

R. B. Laughlin and D. Pines: The theory of everything. *Proc. Nat. Acad. Sci.* **97** (2000) 28–31.

J. Leslie: *Universes*. Routledge, London, 1989.

M. Levine: How insects lose their limbs. *Nature* **415** (2002) 848–949.

M. Livio et al.: The anthropic significance of the existence of an excited state of ^{12}C. *Nature* **340** (1989) 281–4.

J. R. Lucas: *The Conceptual Roots of Mathematics*. Intern. Library of Phil., Routledge, London, 2000.

E. A. Maguire et al.: Navigation-related structural change in the hippocampi of taxi drivers. *Proc. Nat. Acad. Sci.* **97** (2000) 4398–404.

Marzoli et al.: Extensive 200-million-year-old continental flood basalts of the Central Atlantic Magmatic Province. *Science* **284** (1999) 616–18.

J. C. McElwain, D. J. Beerling, F. I. Woodward: Fossil plants and global warming at the Triassic-Jurassic boundary. *Science* **285** (1999) 1386–90.

K. R. Miller: *Finding Darwin's God*. Harper Collins, New York, 1999.

J. Mollon: Worlds of difference. *Nature* **356** (1992) 378–9.

J. D. Mollon: The uses and origins of primate colour vision. pp. 379–96 in *Readings on Color. Vol. 2*, The Science of Color. eds. A. Byrne, D. R. Hilbert, MIT Press, Cambridge, Mass., 1997.

A. Morton, S. P. Stich eds.: *Benacerraf and his Critics*. Blackwell Publ., Cambridge MA, 1996.

D. Mumford et al.: *Indra's Pearls, The Vision of Felix Klein*. Cambridge Univ. Press, Cambridge, England, 2002.

C. D. Murray: Chaotic motion in the Solar System. *Encycl. of the Solar System*, eds. T. Johnson, P. Weissman, L. A. McFadden. Acad. Press, Orlando, 1998.

C. D. Murray, S. F. Dermott: *Solar System Dynamics*. Camb. Univ. Press, Cambridge, 1999.

T. Nagel: Consciousness and objective reality. pp. 63–8 in *The Mind-Body Problem*, eds. R. Warner, T. Szuba, Blackwell Ltd., Oxford, 1994.

R. Newton: *The Truth of Science*. Harvard Univ. Press, Cambridge, MA, 1997.

Okada et al.: Assymmetrical development of bones and soft tissues during eye migration of metamorphosing Japanese flounder, *Paralichthys olivaceus*. *Cell Tissue Res.* **304** (2001) 59–66.

Olsen P. E.: Giant lava flows, mass extinctions and mantle flows. *Science* **284** (1999) 604–5.

R. Omnes: Consistent interpretations of quantum mechanics. *Rev. Mod. Phys.* **64** (1992) 339–358.

D. Papineau: Why supervenience? *Analysis* **50** (1990) 66–71.

D. Papineau: The reason why. *Analysis* **51** (1991) 37–40.

D. Papineau: The rise of physicalism. pp. 3–36 in *Physicalism and its Discontents* eds. Carl Gillett, Barry Loewer, Camb. Univ. Press, 2001.

R. Penrose: *The Emperor's New Mind*. Oxford Univ. Press, Oxford, 1989.

R. Penrose: *Shadows of the Mind*. Oxford Univ. Press, Oxford, 1994.

R. Penrose: Must mathematical physics be reductionist? Ch. 2 in *Nature's Imagination, the Frontiers of Scientific Vision*, ed. J. Cornwall Oxford Univ. Press, Oxford, 1995.

R. Penrose: Beyond the Doubting of a Shadow. *Psyche* 2(**23**) January 1996, p. 31.

M. Perutz: *Is Science Necessary?* Barrie and Jenkins, London, 1989.

Phillips, Lord (chair): *The BSE Inquiry Report*, Crown Copyright, 2000.

Pinker: *The Language Instinct*. Penguin Books Ltd., London, 1995.

H. Poincaré: *Science and Hypothesis*. Dover Publ. Inc., New York, 1952. French original, 1902.

J. Polkinghorne: *Beyond Science*. Cambridge Univ. Press, Cambridge, 1996.

J. Polkinghorne: *Science and Theology*. SPCK/Fortress Press, London, 1998.

K. Popper: *The Logic of Scientific Discovery*. Hutchinson and Co., London, 1959. English translation of German original of 1934.

K. R. Popper: *The Open Universe, An Argument for Indeterminism*. Rowman and Littlefield, Totowa, New Jersey, 1982.

S. Psillos: The present state of the scientific realism debate. *Brit. J. Phil. Sci.* **51** (2000) 705–28.

H. Putnam: The 'corroboration' of theories. pp. 122–38 in *The Philosophy of Science*, eds. R. Boyd et al., MIT Press, Cambridge, MA, 1991.

Y. Rav: Philosophical Problems of Mathematics in the Light of Evolutionary Epistemology. pp. 80–109 in *Math Worlds* eds. Sal Restivo, Jean Paul Van Bendegen, Roland Fischer, State University of New York Press, Albany, 1993.

M. Rees: *New Perspectives in Astrophysical Cosmology*. Camb. Univ. Press, Cambridge, 1995.

M. Rees: *Before the Beginning*. Simon and Schuster, London, 1997.

M. Rees: *Just Six Numbers*. Weidenfeld and Nicolson, London, 1999.

C. A. Ronan: *The Shorter Science and Civilisation in China. An Abridgement of Joseph Needham's Original Text. Vol. 1*, Camb. Univ. Press, Cambridge, 1978.

S. Rose: *Lifelines. Biology, Freedom, Determinism*. Allen Lane, The Penguin Press, London, 1997.

D. Ruelle: *Conversations on mathematics with a visitor from outer space*. Preprint, July, 1998.

M. Ruse: *Can a Darwinian be a Christian?* Cambridge Univ. Press, Cambridge, 2001.

G. Ryle: *The Concept of Mind*. Penguin Books, Harmondsworth, 1973.

D. G. Saari and Z. J. Xia: Off to infinity in a finite length of time. *Notices Amer. Math. Soc.* **42** (1995) 538–46.

J. Searle: Minds. Brains and Programs, in *The Behavioural and Brain Sciences*, Cambridge Univ. Press, 1980.

J. Searle: *Minds. Brains and Science. The 1984 Reith Lectures*. Penguin Books, London, 1984.

J. Searle: What's wrong with the philosophy of mind? pp. 277–98 in *The Mind-Body Problem* eds. R. Warner, T. Szuba, Blackwell Ltd., Oxford, 1994.

T. Seeley, S. Buhrmann: Group decision making in a swarm of honey bees. *Behavioural Ecology and Sociobiology* **45** (1999) 19–31.

S. Shapiro: *Thinking about Mathematics. The Philosophy of Mathematics.* Oxford Univ. Press, Oxford, 2000.

R. N. Shepard: The perceptual organisation of colors: an adaptation to regularities of the terrestrial world? pp. 311–56 in *Readings on Color. Vol. 2, The Science of Color*, eds. A. Byrne, D. R. Hilbert, MIT Press, Cambridge, Mass., 1997.

M. Stöltzner: What Lakatos could teach the mathematical physicist, in *Appraising Lakatos. Mathematics, Methodology and the Man.* eds. G. Kampis et al. Dordrecht. Kluwer, 2001.

J. G. Taylor: The central role of the parietal lobes in consciousness. *Consciousness and Cognition* **10** (2001) 379–417.

J. F. Traub, A. G. Werschulz: *Complexity and Information.* Camb. Univ. Press, Cambridge, 1998.

A. Turing: Computing Machinery and Intelligence. *Mind* **59** (1950) 433–60.

T. Tymoczko: The four colour problem and its philosophical significance. pp. 243–66 in *New Directions in the Philosophy of Mathematics*, ed. T. Tymoczko, Princeton University Press, 1998.

K. Visscher, S. Camazine: Collective decisions and cognition in bees. *Nature* **397** (1999) 400.

H. Wang: On 'computabilism' and physicalism: some subproblems. Ch. 11 in *Nature's Imagination, the Frontiers of Scientific Vision*, ed. J. Cornwall, Oxford Univ. Press, Oxford, 1995.

R. Warner, T. Szuba: *The Mind-Body Problem.* Blackwell Ltd., Oxford, 1994.

A. Wegener: *The Origin of Continents and Oceans.* Third edition, Metheun and Co., London, 1924.

S. Weinberg: *Dreams of a Final Theory.* Hutchinson Radius, London, 1993.

J. Weiner: *The Beak of the Finch.* Alfred A Knopf, New York, 1994.

D. Zeilberger: Theorems for a price: Tomorrow's semi-rigorous mathematical culture. *Notices Amer. Math. Soc.* (1993) 978–81.

J. Ziman: *Of One Mind: The Collectivization of Science.* Amer. Inst. Phys., Woodbury, NY, 1995.

Index